Ariacutty Jayendran

# Englisch
# für Maschinenbauer

## Aus dem Programm
## Grundlagen Maschinenbau

**vieweg**

Ariacutty Jayendran

# Englisch für Maschinenbauer

## Lehr- und Arbeitsbuch

6., erweiterte Auflage

Mit 90 Abbildungen

Viewegs Fachbücher der Technik

Bibliografische Information Der Deutschen Nationalbibliothek
Die Deutsche Nationalbibliothek verzeichnet diese Publikation in der
Deutschen Nationalbibliografie; detaillierte bibliografische Daten sind im Internet
über <http://dnb.d-nb.de> abrufbar.

1. Auflage 1994
2., verbesserte Auflage 1997
3., überarbeitete und erweiterte Auflage 2000
4., durchgesehene Auflage Juni 2002
5., korrigierte und erweiterte Auflage April 2004
6., erweiterte Auflage 2007

Lektorat: Thomas Zipsner

Der Vieweg Verlag ist ein Unternehmen von Springer Science+Business Media.
www.vieweg.de

Umschlaggestaltung: Ulrike Weigel, www.CorporateDesignGroup.de

Gedruckt auf säurefreiem und chlorfrei gebleichtem Papier.

ISBN 978-3-8348-0131-9

## Vorwort zur sechsten Auflage

Für die sechste Auflage sind folgende Änderungen vorgenommen worden:

- Das Buch wird in Zukunft in zwei Teile geteilt. Der erste Teil enthält die 24 Kapitel mit Übungen und Lösungen.
- Der zweite Teil enthält zwei neue Texte in Teil zwei, nämlich "Basics of Robots" und "A visit to an industrial fair". Die Texte in dem ersten Teil sind sehr kurz geschrieben, so dass genug Platz für die Übungen und Lösungen vorhanden ist. Im zweiten Teil gibt es längere Texte aber keine Übungen. Nachdem die Studenten den ersten Teil durchgearbeitet haben, sollten sie in der Lage sein, längere Texte ohne Übungen zu lesen und zu verstehen.
- Das Englisch/Deutsch Vokabular am Ende des Buches ist über-arbeitet und erweitert worden.
- Viele Leser haben ein Deutsch/Englisch Vokabular gewünscht und in dieser Auflage ist ein Deutsch/Englisch Vokabular eingefügt worden.
- Der Index ist erweitert worden.
- Die Appendices I – III sind korrigiert und erweitert worden.

Entstanden ist dieses Buch aus einem sechsmonatigen technischen Englisch Kurs für Fachhochschulstudenten des Maschinenbaus. Obwohl bereits eine Vielzahl von Lehrbüchern über technisches Englisch existiert, ist offensichtlich kein Werk erhältlich, das speziell auf die Belange von Maschinenbaustudenten abgestimmt ist.

Der erste Teil des Buches besteht aus vierundzwanzig Lehreinheiten, wobei jede ein anderes Thema aus dem Bereich Maschinenbau oder Betriebswirtschaft behandelt. In jeder Lehreinheit wird in einem Text ein bestimmter Sachverhalt erläutert, an den sich ein Vokabelglossar sowie drei Übungen anschließen. Ein Glossar aller Vokabeln befindet sich am Ende des Buches. Eine Vielzahl von Diagrammen und Zeichnungen wurde eingefügt, um die Sachverhalte zu veranschaulichen und das Verständnis zu erhöhen.

Dieses Buch ist für jeden geeignet, der über ein Grundwissen an Englisch verfügt, Schul-Englisch sollte völlig ausreichend sein. Es wird nicht der Versuch unternommen, Grammatik zu thematisieren; das Buch wendet sich nicht an absolute Anfänger in Englisch.

Ich empfehle allen Lesern die zusätzliche Verwendung eines elektronischen Wörterbuchs wie das **"Cambridge Advanced Learner's Dictionary on CD-ROM"** (Cambridge University Press). Diese sind nicht teuer und beinhalten englisches und amerikanisches Fachvokabular mit Audiounterstützung.

Das vorliegende Buch eignet sich sowohl für den Unterricht in technischem Englisch, als auch für das Selbststudium und als sprachliche Referenz. Der Umfang der nun vorliegenden sechsten Auflage erlaubt es, das Buch entweder unter Auswahl geeigneter Kapitel als Basis für einen einsemestrigen Kurs oder als Basis für einen zweisemestrigen Kurs zu verwenden.

Ich veröffentliche diese Texte in der Hoffnung, dass sie Studenten, Ingenieuren und Technikern in Praxis und Ausbildung eine Hilfe sind. Ich habe mich bemüht, den Text einfach zu gestalten, indem ich bei der Sprachwahl darauf geachtet habe, kurze Sätze und nicht zu komplizierte Konstruktionen zu verwenden, um das Verständnis zu erleichtern.

Ich möchte an dieser Stelle zwei meiner Studenten, Dirk Hüseken und Ingo Klaussen, sowie meinen Söhnen Rajah und Amir meinen Dank ausprechen, die mir beim Erstellen des Buchs behilflich waren. Ich möchte mich auch bei Herrn Thomas Zipsner vom Vieweg Verlag bedanken für neue Vorschläge und Anregungen für die 6. Auflage.

Wetter, im August 2006 *Ariacutty Jayendran*

Nach dem tragischen Tod meines Vaters kann er nun leider nicht mehr die neue vorliegende 6. Auflage seines Buches miterleben. Ich möchte mich bei allen Lesern für das große Interesse an diesem Buch bedanken, welches ich von nun an im Sinne meines Vaters weiterführen werde.

Wetter, im August 2007 *Rajah Jayendran*

# Contents     **Part 1**

## **Part 2**

# 1 Engineering materials

Manufacturing industry uses a large variety of materials for the manufacture of goods today. Among the most important of these are *metals, alloys*, and *plastics*. Different kinds of metals and alloys are required for different purposes, and the diagram given below shows some of the metals and alloys that are in common use.

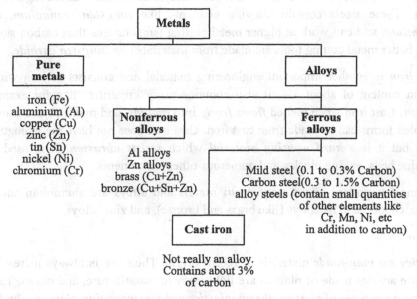

The materials which are probably the most useful are iron and its alloys. Very pure iron is called *wrought iron*. It has the advantage that it can be easily bent and formed into various shapes. It also does not rust easily. However, it lacks the strength required for many applications.

*Steel* is a much stronger material than iron. It is an *alloy* composed of *iron* and *a small amount of carbon*,the carbon content being less than two percent. Sometimes small quantities of other metals are also introduced into steel, thus producing a range of different *alloy steels*, each one being suitable for a different application.The commonest type of steel is *mild steel*, which contains about 0.2 %

of carbon. It is used for many purposes, such as for reinforcing concrete, for the making of car bodies, etc. Mild steel *cannot be hardened*.

*Carbon steel* has a higher carbon content than mild steel and *can be hardened*. It is used for making small tools like screwdrivers, scissors, etc., and also for larger objects like springs and axles. Carbon steel tools *lose their hardness* easily when they become heated during a *metal cutting* process. For this reason, tools made from carbon steel are unsuitable for use in metal cutting machines such as lathes.

Metal cutting tools are often made of special alloy steels usually called *high speed steels*. These steels contain alloying elements like *tungsten*, *vanadium*, and *chromium*, and can work at higher metal cutting temperatures than carbon steels. Even better metal cutting tools are made from materials like *tungsten carbide*.

*Cast iron* is another important engineering material and consists of iron with a carbon content of about 3%. It also contains many impurities like for example sulphur. Cast iron when melted *flows freely* into *moulds*, and many objects having complex forms can be made from cast iron. Cast iron does not have the strength of steel, but it is a *hard wearing material* which *resists abrasion*. It is used for machine beds, engine blocks, and numerous other components.

The most important of the *nonferrous metals* and *alloys* are aluminium and its alloys, copper and its alloys (like brass and bronze), and zinc alloys.

## Plastics

*Plastics* are *man-made* materials of recent origin. Their use is always increasing, because articles made of plastics are light, easy to manufacture, and cheap. There are two main types of plastics, *thermoplastics* and *thermosetting plastics*. The first type includes nylon, polyethylene and polyvinylchloride. These become soft when heated, and can be moulded into different shapes. On cooling, they become hard again. Such plastics can be remoulded and reused repeatedly. They are used for making articles like water pipes, combs, etc. *Thermosetting plastics* however, can be heated and moulded only once. They are usually harder than thermoplastics, and are used to mould objects like electrical switches, cups, plates, etc.

## Improved properties obtained by combining materials

When a single material does not have the required properties, a combination of materials may be able to meet the requirements.

- *Alloys* are composed of several elements, and have special properties which the individual elements do not usually have.

- *Composite materials* like *resins reinforced* with *fibre glass*, have the advantages of high strength and easy mouldability at room temperatures. Other common composite materials are *reinforced concrete* and *wood* (a natural composite material made of cellulose and lignin).

- The addition of *small amounts* of *impurities* can cause big changes in material properties. *Free-cutting steel* which is a special steel that can be machined easily, is produced by adding lead to ordinary steel. This is not a steel alloy, because lead does not form an alloy with steel. *Modern electronic devices* are made from *two types* of *silicon, n* and *p*. These two important materials are produced by adding small amounts of impurities to very pure silicon.

---

# Vocabulary

| | | | |
|---|---|---|---|
| abrasion | Abrieb *m*, Abnutzung *f* | mild steel | Baustahl *m* |
| advantage | Vorteil *m* | mould | formen, bilden *v* |
| alloy | Legierung *f* | mould | Gießform *f*, Form *f* |
| carbon steel | Kohlenstoffstahl *m* | origin | Ursprung *m* |
| cast iron | Gußeisen *n* | probably | wahrscheinlich *adv* |
| composed of | zusammengesetzt aus *v* | property | Eigenschaft *f* |
| composite | zusammengesetzt *adj* | purpose | Zweck *m* |
| contain | enthalten *v* | range | Reihe *f* |
| content | Inhalt *m* | reinforce | verstärken *v* |
| ferrous | eisenhaltig *adj* | reinforced concrete | Eisenbeton *m* |
| free-cutting steel | Automatenstahl *m* | resin | Harz *n* |
| hard wearing | widerstandsfähig *adj* | strength | Festigkeit *f*, Stärke *f* |
| increase | zunehmen *v* | suitable | geeignet *adj* |
| introduce | einführen *v* | wear | Verschleiß *m* |
| lack | mangeln, fehlen *v* | wrought iron | Schweißstahl *m* |

# Exercises I

## 1. Answer the following questions:

a) What advantages does wrought iron have over other types of iron and steel ?

b) What is mild steel, and for what purposes is it used ?

c) What is carbon steel, and why are tools made from carbon steel unsuitable for use in metal cutting machines ?

d) In what way are alloy steels different from carbon steel ?

e) What is cast iron, and what advantages does it have as an engineering material ?

f) Which nonferrous metals and alloys are commonly used as engineering materials ?

g) What advantages do goods made from plastics have over goods made from other materials ?

h) What are the relative advantages of thermoplastics and thermosetting plastics ?

i) Give an example of a composite material, and state what advantages it has over other materials.

j) How are the two types of silicon n and p, manufactured ?

## 2. Fill in the gaps in the following sentences:

a) Different _____ of metals and _____ are required for different _____ .

b) Wrought iron has the _____ of being _____ bent into _____ shapes.

c) Steel is an alloy _____ of iron and a small _____ of carbon.

d) Carbon steel can be _____ , and is _____ to make small _____ like scissors and screwdrivers.

e) Metal cutting tools are made of _____ steels called high _____ steels.

f) Tools made of alloy steels can work at _____ than tools ____ of _____ steel.

g) Cast iron when ____ flows ____ into moulds.

h) Thermoplastics ____ soft when ____ and can be ____ into different shapes.

i) Composite materials can be made from resins ____ with ____.

j) Mild steel can be used for _____ concrete and for making ____ bodies.

## 3. Translate into English:

a) Metalle, Legierungen und Kunststoffe gehören zu den wichtigsten Materialien, die heute in der verarbeitenden Industrie eingesetzt werden. Unterschiedliche Anwendungen erfordern unterschiedliche Metalle und Metalllegierungen.

b) Baustahl, der ca. 0,2% Kohlenstoff enthält, ist die am weitesten verbreitete Art von Stahl und wird für viele Zwecke eingesetzt, wie Verstärkung von Beton, Herstellung von Wassertanks, etc. Baustahl kann im Gegensatz zu Stahl, welcher einen höheren Kohlenstoffanteil enthält, nicht gehärtet werden.

c) Gusseisen ist ein sehr nützliches Material, da es im flüssigen Zustand leicht in Gießformen fließt. Auch komplexe Gegenstände können aus Gusseisen hergestellt werden. Gusseisen ist ein widerstandsfähiges Material, das Abrieb widersteht.

d) Kunststoffe sind von zunehmender Bedeutung, da Artikel aus Kunststoff einfach herzustellen, preiswert und leicht sind. Einige Kunststoffe werden bei Erwärmung weich und bei Erkaltung wieder hart. Solche Kunststoffe können wiederholt verarbeitet werden.

e) Verbundmaterialien, die aus Harzen und Glasfasern hergestellt werden, kommen dort zum Einsatz, wo hohe Festigkeit und einfache Verformbarkeit bei Raumtemperatur zu den erforderlichen Eigenschaften gehören.

## 2  Properties of materials

Before appropriate materials for the manufacture of a given product are chosen, a *careful evaluation* of the properties of all *available materials* is usually advisable. Among the properties which may need to be considered are mechanical properties, physical properties, and chemical properties.

### Mechanical properties

Foremost among the properties that should be considered are mechanical properties like *strength, hardness, ductility, and toughness*. A very important way of assessing some of the mechanical properties of a material is by means of a Young's modulus experiment. A brief description of such an experiment is given below.

### Young's modulus experiment for a uniform metal specimen

In this experiment, a piece of material of uniform cross-section is subjected to a *tensional load* which increases the length of the specimen. The load is gradually increased from zero to a value at which the *specimen breaks*. The following physical quantities are used to study the behaviour of the specimen quantitatively.

> **Longtudinal stress = Tensile force /cross-sectional area    ( N/m$^2$ )**
>
> **Longitudinal strain = Increase in length/original length    ( No units )**
>
> **Young's modulus E = Stress / Strain   ( N/m$^2$ )**

If a *graph* between *stress* and *strain* is plotted (after the experiment has been completed), a curve similar to that in Fig. 2.1 is obtained. *In the region OA, the strain is proportional to the stress.* If the load is removed at this stage, the specimen will revert to its *original shape* and *size.* The behaviour of the material in this region is said to be *elastic,* and the change in shape it undergoes when loaded is called *elastic deformation.* The stress corresponding to the point A where the proportional region ends, is called the *yield strength.*

At higher values of stress beyond A, *the slope* of the curve *changes*, and the deformation is *no longer proportional* and *elastic.*The specimen does not return to its original size and shape when the load is removed, and is said to have undergone *plastic deformation*.

In the region BC, the specimen *elongates permanently* and at the same time its *cross-section* becomes *reduced uniformly* along its length. The material becomes stronger and is said to be *strain hardened* or *work hardened* .

Beyond the point C, the cross-section of the specimen *decreases* only at its *weakest point*. This process is called *necking* (see Fig. 2.1 (b)), and the load that can be sustained decreases. Finally at point C *fracture occurs*. The stress at the maximum load (point B) is called the *ultimate tensile strength* (**UTS**) or the tensile strength (TS). The stress at point C is called the *breaking strength*.

**Elastic and plastic deformation in engineering applications**

Elastic and plastic deformation are both useful in engineering applications. In *structural applications* like columns and machine frames, loads must be kept well within the *elastic range* so that *no permanent deformation occurs*. In applications like the bending, stretching, and the deep drawing of metal, *plastic deformation has to take place* if the shape of the metal is to be changed.

**Some important mechanical properties**

- *Strength* refers to the ability of a material to *resist tensional* or *compressional stresses*. The yield strength and the ultimate tensile strength of a material are important quantities in engineering design.

- *Elasticity* is the ability of a material to return to its original shape and dimensions after load has been removed.

- *Plasticity* is the ability of a material to be permanently deformed without breaking.

- *Ductility* refers to the property of a metal which allows it to be *plastically deformed* and to be *drawn into wire* without breaking.

- *Toughness* refers to the ability of a material to *withstand bending* (or the application of shear stresses) *without breaking*. Copper is extremely tough, while cast iron is extremely brittle.

- **Brittleness** refers to a property possessed by some materials like cast iron. Such materials *can only be deformed elastically*, and *break easily* when they are *subjected to plastic deformation.* Brittleness is the opposite of toughness.

- **Hardness** refers to the ability of a material *to resist abrasion* or *indentation*. The hardness of a material can be measured by a hardness test like a **Brinell test.** Hardness is often a surface property.

- **Impact properties** - Some materials which are usually tough and ductile, suffer *fracture* when they are in the form of a *notched specimen* and subjected to a sudden load or *impact.* Standard tests like an *Izod test* are available to test the impact properties of a material.

- **Fatigue** - When a material is subjected to constant loads well below the yield strength of the material, permanent deformation does not normally occur. However, *repeated application* of *even small loads* which are well below the value of the yield strength, can cause *cracks to appear*. These cracks can gradually become bigger, and finally lead to *fatigue fracture*.

- **Creep** - Metals subjected to stresses which are below the ultimate tensile stress for *long periods of time,* can undergo gradual extension and ultimately break. This phenomenon usually occurs at high temperatures and is called creep.

- **Properties** which *facilitate manufacture* like easy machinability, free flow of the molten metal during casting, and good welding properties also need to be considered when choosing a material for a particular application.

# Vocabulary

| | | | |
|---|---|---|---|
| ability | Fähigkeit *f* | frame | Gestell *n* |
| advantage | Vorteil *m* | gradually | allmählich *adv* |
| appropriate | geeignet *adj* | hardness | Härte *f* |
| assess | bewerten *v* | indentation | Eindruck *m* |
| behaviour | Verhalten *n* | impact | Stoß *m*, Schlag *m* |
| bend | biegen *v* | load | Last *f* |
| breaking strength | Bruchfestigkeit *f* | necking | Querschnitt-verminderung *f* |
| brittleness | Sprödigkeit *f* | measure | messen *v* |
| carry out | durchführen *v* | notch | Kerbe *f* |
| change | Änderung *f* | phenomenon | Phänomen *m*, Erscheinung *f* |
| choose | wählen, aussuchen *v* | plot | graphisch darstellen *v* |
| column | Säule *f* | remove | entfernern *v* |
| consider | erwägen *v* | revert | zurückkehren *v* |
| corresponding | entsprechend *adj* | shear stress | Scherbeanspruchung *f* |
| crack | Riss *m* | slope | Steigung *f* |
| creep | kriechen *v* | specimen | Versuchsgegenstand *m*, Exemplar *n* |
| cross-section | Querschnitt *m* | strain | Dehnung *f* |
| decrease | sich vermindern *v* | strain harden | kalthärten *v* |
| deformation | Verformung *f* | stress | Spannung *f* |
| ductility | Dehnbarkeit *f* | subject to | belasten mit *v* |
| elongation | Verlängerung *f* | tensional load | Zugbelastung *f* |
| evaluate | berechnen *v* | toughness | Zähigkeit *f* |
| facilitate | ermöglichen *v* | undergo | erfahren *v* |
| fatigue strength | Dauerfestigkeit *f* | ultimate tensile strength | Zugfestigkeit *f* |
| fracture | Bruch *m* | work harden | kalthärten *v* |

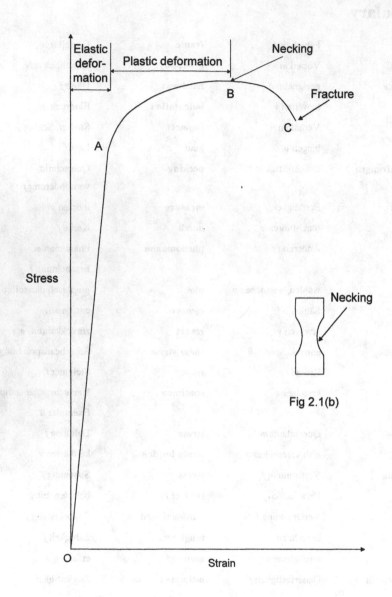

Fig 2.1 (a) Stress versus strain curve for a uniform metal specimen

# Exercises II

## 1. Answer the following questions:

a) Give the names of four of the most important mechanical properties of a material.

b) State the name of an experiment from which some of the most important mechanical properties of a material can be evaluated.

c) How are (longitudinal) stress and strain defined ?

d) What is the relationship between stress and strain in the elastic region ?

e) Explain the meaning of the term necking.

f) What kinds of loads should be used for structural applications in engineering ?

g) What do you understand by ductility ?

h) What happens to a brittle material when it undergoes plastic deformation ?

i) Explain what you understand by the expression, toughness of a material.

j) What do you understand by the expression, hardness of a material.

## 2. Fill in the gaps in the following sentences:

a) Foremost _____ the properties that should be _____ are mechanical properties.

b) The load is _____ from zero until the _____ breaks.

c) If the _____ is removed, the specimen will revert to its _____ state.

d) The stress at the _____ load is called the ultimate _____.

e) Elastic and plastic _____ are both useful in engineering _____.

f) Toughness refers to _____ of a material to _____ bending.

g) Ductility _____ to the property of of a material which allows it _____ into wire.

h) Creep refers to a ____ which usually occurs at ____.

i) Properties which ____ manufacture need also to be ____.

j) Hardness refers to the ____ of a material to ____ indentation.

## 3. Translate into English:

a) Eine wichtige Methode, die hauptsächlichen mechanischen Eigenschaften eines Materials zu beurteilen, ist die Durchführung eines Experiments zur Begutachtung seiner Längenänderung, wenn es einer Zugbelastung ausgesetzt wird.

b) Bei höheren Spannungswerten jenseits von A, ändert sich die Steigung der Kurve und die Verformung ist nicht mehr proportional und elastisch. Der Versuchsgegenstand kehrt nicht in seine ursprüngliche Form und Größe zurück, wenn die Belastung entfernt wird.

c) Bei strukturellen Anwendungen wie Säulen oder Rahmen für Maschinen, müssen die Lasten weit im elastischen Bereich gehalten werden, so dass keine dauerhafte Verformung auftritt.

d) Die Dehnbarkeit bezieht sich auf die Eigenschaft eines Metalls, die es erlaubt es zu einem Draht zu ziehen, ohne dass er zerbricht. Ein dehnbares Material muss sowohl Härte als auch Plastizität aufweisen. Blei ist beispielsweise dehnbar, aber lässt sich schwer zu einem Draht ziehen, weil die Härte gering ist.

e) Die Härte bezieht sich auf die Fähigkeit eines Materials, Abrieb und mechanischen Eindrücken zu widerstehen. Die Härte eines Materials kann mittels eines Härtetests wie dem Brinell-Test gemessen werden.

# 3  Hand Tools and workshop equipment

The efficient use of machines, computers, and automation has enabled modern manufacturing industry to reach such an advanced state, that the need for hand tools has been almost eliminated. However, hand tools are still required for many purposes in industry, like the making of *prototypes* and *models*, and for the *repair* and *maintenance* of equipment. Some of the most common hand tools are shown in Fig 3.2 and a brief description of their use is given below.

A *hammer* is a multipurpose tool which can be used for hammering nails, for shaping and forming sheet metal, for tapping together closely fitting parts, and so on.

*Punches* are of many types. A *centre punch* is used to mark a specific point on a piece of metal before drilling. *Drift punches* are used to align two or more pieces of metal which are to be joined together with bolts or rivets. A *pin punch* is used to drive in or remove straight pins, tapered pins, and keys.

*Screwdrivers* are common tools which are used to turn screws. An *offset screwdriver* is very useful for turning screws in awkward places. *Pliers* are of many different types and are used for many purposes, like holding, gripping, and turning. A *spanner* or a *wrench* is a tool used for turning nuts and bolts. There are many types of spanners, like the *double-ended spanner*, *ring spanner*, *socket spanner*, etc.

The cutting of metals by hand is usually done using a *hacksaw*. *Files* are used to remove metal from the surface of an object until it reaches the desired shape and size. Small quantities of metal on *high spots* of a metal surface can be removed by using a *scraper*.

*Chisels* are used together with a hammer to chip away unwanted parts of an object made of metal, wood, or stone. Chisels and all cutting tools need to have a definite shape, so that they can work efficiently without becoming too hot. The important angles, *rake angle* and *clearance angle*, are shown in Fig 3.1.

*Taps* are used for cutting internal screwthreads, and *dies* are used for cutting external screwthreads. Fig 3.2 (b) shows a tap and a *tap wrench* which holds the tap when it is being turned to cut the thread. It also shows a *circular split die* and a *die stock* (or holder for the die). In addition to the hand tools mentioned, some of the other types of equipment required in a workshop are briefly described below.

### Holding and clamping devices

The most important of these is a *general purpose vice*. The object is held firmly between the jaws of the vice while work is being done on it. Other devices are *toolmaker's clamps, vee-blocks, and angle blocks.* Vee-blocks are used to hold cylindrical rods or pipes while work is being done on them. Angle blocks have two surfaces which are perpendicular to each other, and are useful when the position of the work has to be changed by 90°. Some of these devices are shown in Fig 3.3 (a).

### Marking out equipment

It is often necessary to scribe *points, lines, circles*, and other shapes on objects or castings before working on them. *Marking-out equipment* is used for this purpose. Some of the types of equipment commonly used are shown in Fig 3.3 (b).

### Basic measuring devices

Measuring devices are required for the marking-out process, and also for checking the dimensions of a work piece. A *vernier caliper* can measure internal and external dimensions to an accuracy of 1/100 cm. Micrometers are basically more accurate than vernier calipers, and can be used to make internal, external, and depth measurements. *Inside and outside calipers* are often used by machine tool operators to check the dimensions of an object during the machining process. *Try squares* are used to check if two surfaces are perpendicular to each other.

### The use of gauges for comparison and measurement (see also Appendix III)

A *mechanical gauge* (American: gage) is usually a hardened steel device which is used to check the *dimensions, limits, tolerances, form,* etc. of an object.

*Slip gauges* (American: gage blocks) are very precisely made hardened steel blocks, each having a different length. They are available in sets and are used as length standards. Any length can be obtained by *wringing* (combining) several gauges in a set. *Angle gauges* are similar and can be combined to give any angle.

*Plug and ring gauges* can be used for the measurement of diameters. These are of the *go* or *not-go* type. Any object with dimensions outside the given tolerance (see chap 13) is rejected by these gauges. *Contour gauges (templates)* are used to check forms visually. Other types of gauges like *electronic, pneumatic, dial, strain,* etc. are also widely used.

## Vocabulary

| | | | |
|---|---|---|---|
| accurate | genau *adj* | micrometer | Messchraube *f* |
| acquire | erwerben *v* | multipurpose | Mehrzweck- |
| advanced | fortgeschritten *adj* | nut | Mutter (Schrauben) *f* |
| although | obwohl *cj* | offset screw driver | Winkelschrauben-dreher *m* |
| awkward | schwer zugänglich *adj* | part | Teil *n* |
| bolt | Schraube *f* (mit Mutter) | pipe | Rohr *n* |
| chisel | Meißel *m* | pliers | Zange *f* |
| clamp | Klemme *f* | prototype | Muster *n* |
| description | Beschreibung *f* | punch | Stanzwerkzeug *n* |
| device | Gerät *n* | purpose | Zweck *m* |
| dial gauge | Messuhr *f* | rivet | Niete *f* |
| die | Schneideisen *n* | rod | Rundstab m |
| eliminate | beseitigen *v* | rough | rau *adj* |
| enable | ermöglichen *v* | screw | Schraube *f* |
| equipment | Ausstattung , Einrichtung *f* | spanner | Steckschlüssel *m* |
| experience | Erfahrung *f* | socket spanner | Schraubenschlüssel *m* |
| fixture | Vorrichtung *f*, feste Anlage *f* | sufficient | genug, ausreichend *adj* |
| grip | greifen *v* | tap | Gewindebohrer *m* |
| jig | Vorrichtung *f* | taper | Verjüngung *f* |
| key | Keil *m* | turn | drehen *v* |
| knowledge | Kenntnis *f*, Wissen n | vernier caliper | Messschieber *m* |
| maintenance | Instandhaltung *f* | vice | Schraubstock *m* |

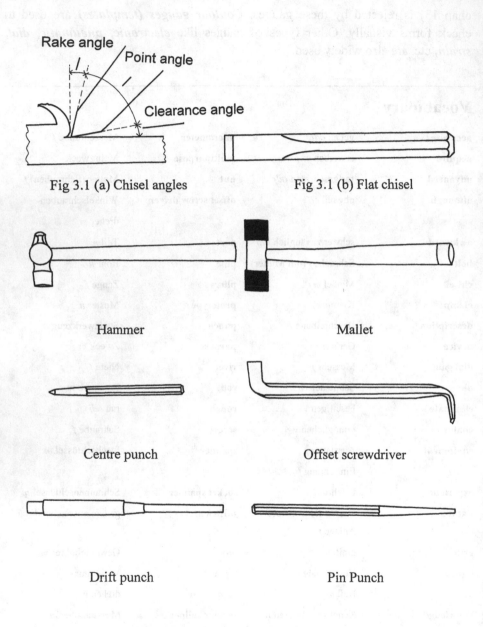

Fig 3.1 (a) Chisel angles

Fig 3.1 (b) Flat chisel

Hammer

Mallet

Centre punch

Offset screwdriver

Drift punch

Pin Punch

Fig 3.2 (a) Some hand tools

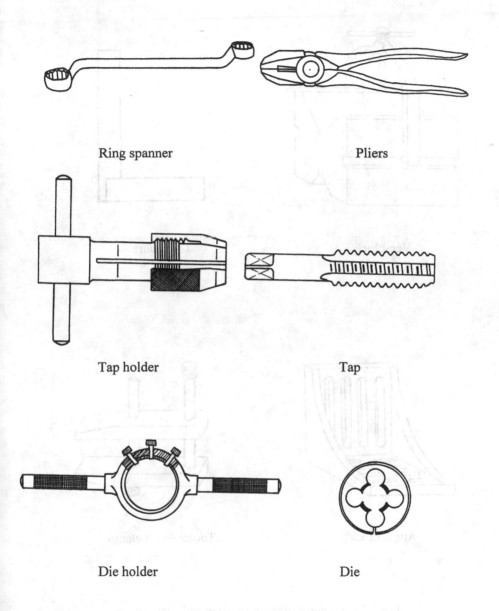

Ring spanner

Pliers

Tap holder

Tap

Die holder

Die

Fig 3.2 (b) Some hand tools

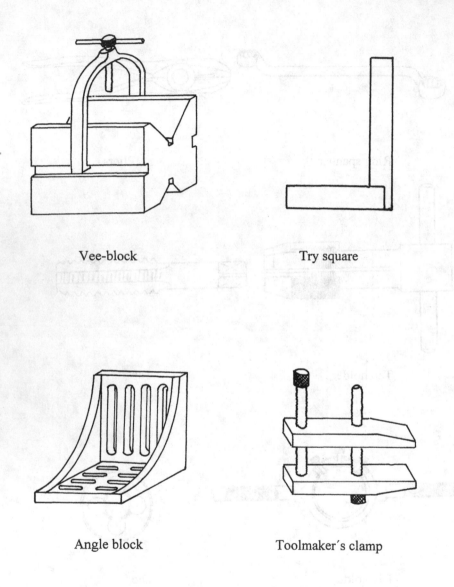

Vee-block

Try square

Angle block

Toolmaker´s clamp

Fig 3.3 (a) Some clamping devices and a try square

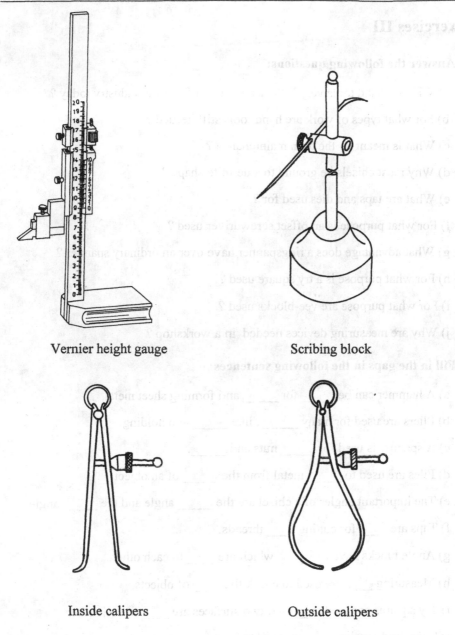

Vernier height gauge                    Scribing block

Inside calipers                         Outside calipers

Fig 3.3 (b) Some marking-out and measuring equipment

# Exercises III

## 1. Answer the following questions:

a) Why are hand tools very little used in manufacturing industry today ?

b) For what types of work are hand tools still needed ?

c) What is meant by the term maintenance ?

d) Why must chisels be ground to a definite shape ?

e) What are taps and dies used for ?

f) For what purpose is an offset screwdriver used ?

g) What advantage does a ringspanner have over an ordinary spanner ?

h) For what purpose is a try square used ?

i) For what purpose are vee-blocks used ?

j) Why are measuring devices needed in a workshop ?

## 2. Fill in the gaps in the following sentences:

a) A hammer can be _____ for _____ and forming sheet metal.

b) Pliers are used for many _____ , like _____ and holding.

c) A spanner is used for _____ nuts and _____.

d) Files are used to _____ metal from the _____ of an object.

e) The important angles of a chisel are the _____ angle and the _____ angle.

f) Taps are _____ for cutting _____ threads.

g) Angle blocks have two _____ which are _____ to each other.

h) Measuring _____ are used to check the _____ of objects.

i) Try squares are used to _____ if two surfaces are _____ to each other.

j) Inside and outside _____ are used by _____ operators.

## 3. Translate into English:

a) Obwohl die meisten Produkte heutzutage mit automatischen Maschinen hergestellt werden, braucht man Handwerkzeuge für Reparaturarbeiten und für die Instandhaltung von Maschinen und Anlagen.

b) Ein Hammer ist ein Vielzweckwerkzeug. Man kann damit unter anderem Metallblech hämmern, bis es eine bestimmte Form annimmt.

c) Ein abgewinkelter Schraubenzieher ist ein Spezialwerkzeug. Er wird benutzt, wenn man eine Schraube an einer schwer zugänglichen Stelle erreichen will.

d) Es gibt verschiedene Sorten von Schraubenschlüsseln. Wenn man eine Schraube sicher greifen möchte, benutzt man einen Ringschlüssel oder einen Gabelschlüssel.

e) Eine Metallsäge, die man zum Trennen von Metallblech benutzt, nennt man im Englischen "hacksaw". Um ein Metallstück in eine bestimmte Form zu bringen, kann man es mit einem Meißel teilen und dann mit einer Feile bearbeiten.

f) Messinstrumente werden beim Markieren benutzt, sowie bei der Kontrolle der Dimensionen nach Bearbeitung.

g) Ein Anschlagewinkel wird benutzt, um zu kontrollieren, ob zwei Flächen in rechtem Winkel zueinander stehen.

# 4  The joining of metals by mechanical methods

There are many ways in which metal sheets and components can be joined together. One possible way is by the use of mechanical fasteners like rivets, bolts and nuts, and set screws.

*Riveting* is useful when metals sheets have to be joined together permanently. It is particularly useful for joining thin metal sheets, and it is often used on ship's hulls and aircraft fuselages. Fig 4.4 shows two rivets holding two thin sheets of metal together. Countersunk rivets must be used for joints that need to have a flush surface. *Fastening devices* which have a screw thread like nuts and bolts, are used to join metal parts which may have to be taken apart later, for purposes of repair or replacement.

*Nuts and bolts* are used when both sides of the parts to be joined are accessible. If the parts are subject to vibration, an additional component like a *spring washer* or a *lock nut* will be necessary to prevent the nut from coming loose. Fig 4.1 shows the use of a bolt and a nut in a clamping device. When the use of a bolt and a nut is not possible, set screws are used. Fig 4.2 shows the typical use of a set screw. *Set screws* with normal heads can be used, but it is often necessary to use screws with *countersunk heads*. Such screws usually have a hole of *hexagonal form* within the head of the screw. This enables the screw to be tightened efficiently by using a hexagonal key as shown in Fig 4.3.

*Studs* are used for the joining of parts to cast iron components. The tensile strength of cast iron is very low, and excessive tightening of a set screw into a cast iron thread may cause the thread to crumble, thus permanently damaging the casting. The studs are first screwed into the casting, and the tightening done by using mild steel nuts. Any damage done by excessive tightening will be to the stud or the nut, and not to the casting. Studs are used to ensure *gas-tight* and *water-tight joints* in applications where *heavy pressures* are encountered. Fig 4.5 shows the use of studs for holding down a cylinder head on a cylinder block of a motor car engine. The joint between the cylinder and the head must be a temporary one, because it is necessary to remove the head when the engine needs reconditioning.

To make the joint gas and water-tight, a *thin gasket* is used between the metal surfaces. The studs and nuts used are usually made of mild steel.

## Vocabulary

| | | | |
|---|---|---|---|
| accessible | zugänglich *adj.* | nut | Mutter *f* |
| aircraft fuselage | Flugzeugrumpf *m* | often | oft *adv* |
| application | Anwendung *f* | permanent | dauernd *adj* |
| bolt | Schraube *f* mit Mutter | recondition | überholen *v* |
| cast iron | Gusseisen *n* | remove | entfernen, beseitigen *v* |
| casting | Guss *m* | repair | reparieren *v* |
| cause | verursachen *v* | replace | ersetzen *v* |
| component | Bestandteil *m* | rivet | Niete *f* |
| countersunk screw | Senkschraube *f* | set screw | Stellschraube *f* |
| crumble | krümeln *v* | solder | löten *v* |
| cylinderblock | Zylinderblock *m* | ship's hull | Schiffsrumpf *m* |
| damage | schaden *v* | spring washer | Unterlegscheibe *f* |
| encounter | begegnen *v* | strip | abziehen *v* |
| engine | Motor *m* | stud | Stiftschraube *f* |
| ensure | sichern, sicherstellen *v* | take apart | zerlegen, auseinandernehmen *v* |
| excessive | übermäßig *adj* | tensile strength | Zugfestigkeit *f* |
| fasten | befestigen *v* | temporary | vorübergehend *adj* |
| gasket | Dichtung *f* | tighten | festziehen, anziehen *v* |
| hexagonal | sechseckig *adj* | useful | nützlich *adj* |
| necessary | notwendig *adj* | vibration | Schwingung , Vibration *f* |

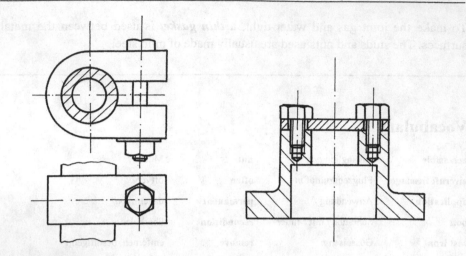

Fig 4.1 Use of bolts and nuts          Fig 4.2 Use of set screws

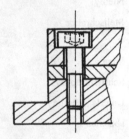          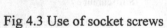          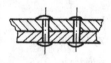

Fig 4.3 Use of socket screws          Fig 4.4 Use of rivets

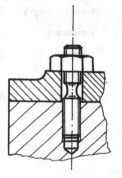

Fig 4.5 Use of studs

# Exercises IV

## 1. Answer the following questions:

a) For what purposes are metal fasteners used ?

b) When are rivets used ?

c) When are fasteners which have a screw thread used ?

d) What can be done to prevent a bolt and a nut from becoming loose ?

e) Why are set screws usually made with a hexagonal hole in their heads ?

f) State one application in which the use of studs is absolutely necessary.

g) What can happen if a bolt is tightened excessively into a cast iron screw thread ?

h) Give an example of an object in which two of its components are held together by studs.

i) What kind of fasteners are used in applications where water-tight joints are required ?

j) When is it not possible to use bolts and nuts as fasteners ?

## 2. Fill in the gaps in the following sentences:

a) Metal components can be _____ together by using _____ fasteners.

b) Riveting is used on ships' _____ and _____ fuselages.

c) Metal parts may also be _____ by fastening _____ which have a screw thread.

d) When both _____ of the _____ to be joined are _____, nuts and bolts may be used.

e) It is often _____ to use set screws with countersunk _____ .

f) Countersunk screws often have a _____ hole in their _____ .

g) Studs are used in _____ where heavy _____ are encountered.

h) Studs are used to join components to cast iron because the _____ strength of cast iron is _____ .

i) The screw _____ in a cast iron _____ may crumble if tightened _____ .

j) It is _____ to remove the head of a motor car engine when the _____ needs _____ .

## 3. Translate into English:

a) Metallteile und Bleche können auf vielfältige Weise verbunden werden. Eine Methode ist die Verwendung von mechanischen Befestigungsvorrichtungen, wie Nieten oder Schrauben und Muttern.

b) Wenn die zu verbindenden Teile von zwei Seiten zugänglich sind, sind Schrauben und Muttern zu verwenden. Ein zusätzliches Teil wie eine Unterlegscheibe könnte nötig sein, falls die Verbindungsteile Vibrationen ausgesetzt sind.

c) Schrauben mit normalen Köpfen können zur Befestigung von Metallkomponenten benutzt werden. Jedoch ist es häufig nötig, Schrauben mit Senkkopf zu verwenden. Solche Schrauben (Inbusschrauben) haben eine hexagonal geformte Vertiefung im Kopf, welche es ermöglicht, die Schraube mit einem hexagonalen Schlüssel (Inbusschlüssel) sehr effektiv anzuziehen.

d) Stiftschrauben werden dort eingesetzt, wo hohe Drücke auftreten und wasser- sowie gasdichte Verbindungen benötigt werden.

e) In Automobilmotoren werden Stiftschrauben eingesetzt, um den Zylinderkopf mit dem Zylinderblock zu verbinden. Die Verbindung zwischen Zylinderblock und -kopf muss temporär sein, weil es zur Überholung des Motors nötig ist, den Zylinderkopf zu entfernen.

# 5 The joining of metals by soldering and welding

*Soldering* is the process of joining two metal objects together by a third soft metal alloy which is called solder. *Solder* is a metal alloy which melts at a lower temperature than the metals being soldered. Two types of solder are commonly used. One is a *soft solder*, which is an alloy of tin and lead. The other is a *hard solder*, which is an alloy of copper and zinc. This is usually called *spelt or silver solder*.

Before soldering two metals, it is very important that their surfaces are first thoroughly cleaned. The material which is used for the purpose of cleaning is called a *flux*. It is used to remove oxides, grease, etc. from the surfaces, allowing the solder to flow freely and unite the surfaces to be joined more firmly. The fluxes usually used in the workshop are *acidic* in character. Common types are zinc chloride, hydrochloric acid, and acid paste. For certain applications, the *corrosive action* of the acid type fluxes must be avoided and a *noncorrosive flux* like *resin* has to be used. This is particularly true of electrical work, where the solder used is usually in the form of a wire with a resin core.

*Brazing* or *hard soldering* differs from soft soldering in that a different kind of solder has to be used, and much more heat has to be applied to reach the melting point of *spelt* (or silver solder) which is about 600°C. *Borax* is normally used as a flux. A brazed joint is much stronger than a soft soldered joint, but can only be used to join articles which can stand the high brazing temperatures without melting.

*Welding* is a process by which metal objects can be joined together forming a very strong and long-lasting bond between them. The main types of welding are shown in the diagram.

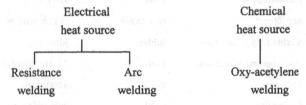

In *oxy-acetylene welding,* a blowpipe through which oxygen and acetylene gases flow in suitable amounts is used. The flame produced can reach maximum temperatures of about 3000°C. The blowpipe heats the two metal parts to be joined, and also a filler rod which is held close to the joint. The filler rod melts, and a small pool of liquid metal is formed at the joint. When the flame is removed, the liquid metal solidifies and a strong joint is formed. Fig 5.1 illustrates oxy-acetylene welding.

The *electric arc welding process* uses a transformer which changes the high voltage low current source at the input terminals, to a low voltage high current source at the output terminals which are used for welding. This process has the advantage that the metal electrode used to strike the arc acts as a filler rod, making the process easier to control than oxy-acetylene welding. Fig 5.2 shows an arc-welding set-up.

In the methods described so far, welding is done by fusion of the filler rod or electrode. In the *resistance welding method*, small portions of the metal objects being welded themselves melt, and when allowed to cool form a very strong joint. In this method *a large current is passed* through the metal surfaces to be joined, the surfaces being held together under mechanical pressure. There is a large *electrical resistance* across the *contact faces*, and this causes a lot of heat to be generated in this region. When the current passes, a small pool of liquid metal is formed and when the current stops, metal pool solidifies forming a strong welded joint. Fig 5.3 and Fig 5.4 show some examples of resistance welding.

## Vocabulary

| | | | |
|---|---|---|---|
| **acid** | Säure *f* | **join** | verbinden *v* |
| **alloy** | Legierung *f* | **oxy-acetylene** | Azetylen-Sauerstoff *m* |
| **arc** | Lichtbogen *m* | **resin** | Harz *n* |
| **braze** | hartlöten *v* | **resistance** | Widerstand *m* |
| **compound** | Verbindung *f*, Gemenge *n* | **solder** | löten *v* |
| **corrosive** | korrodierend, ätzbar *adj.* | **spelt** | Messinghartlot *n* |
| **filler** | Füller *m* | **unite** | vereinigen *v* |
| **flux** | Flussmittel *n* | **weld** | schweißen *v* |

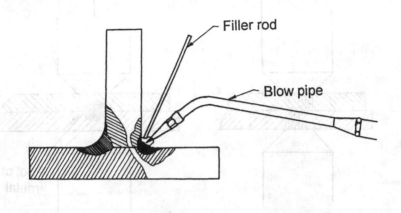

Fig 5.1 Oxy-acetylene welding

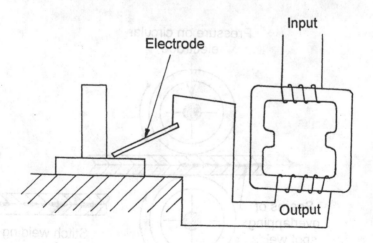

Fig 5.2 Electric arc welding

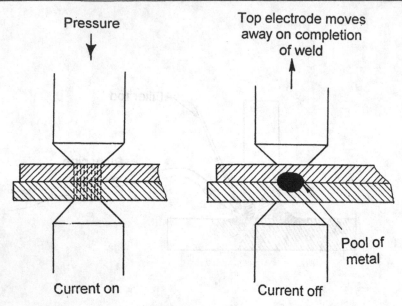

Pressure

Top electrode moves
away on completion
of weld

Pool of
metal

Current on                    Current off

Fig 5.3 Spot welding

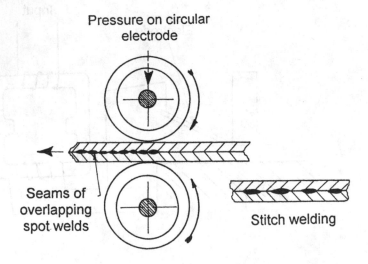

Pressure on circular
electrode

Seams of
overlapping
spot welds

Stitch welding

Fig 5.4 Stitch welding

# Exercises V

## 1. Answer the following questions:

a)  Explain briefly the process of soldering.

b)  What is solder and what types of solder are commonly used ?

c)  What must be done to the surfaces of two metals before they can be soldered ?

d)  What kinds of flux are usually used in a workshop ?

e)  Why is an acid flux unsuitable for electrical work ? What other flux can be used instead ?

f)  In what way is brazing different from soldering ?

g)  What kind of burner or other burning device is used in oxy-acetylene welding ?

h)  What advantage has electric arc welding over oxy-acetylene welding ?

i)  What is the main difference between the resistance welding process and the other methods described here ?

j)  Why is a large amount of heat generated at the contact faces in the resistance welding method ?

## 2. Fill in the gaps in the following sentences:

a)  The process of _____ two metal surfaces by a third soft metal alloy is called _____ .

b)  Hard solder is an alloy of _____ and zinc and is called _____ or silver solder.

c)  Flux removes _____ and grease from the _____ and allows the solder to _____ freely.

d)  Fluxes _____ in the workshop are usually _____ in character, and are _____ .

e) The solder used for electrical work is in the form of a _____ with a _____ .

f) The hard soldering process uses _____ as the solder and _____ as the flux.

g) The two principal types of electrical welding are _____ welding and _____ welding.

h) In oxy-acetylene welding a _____ through which oxygen and acetylene _____ flow through is used.

i) The filler rod melts and a small amount of _____ is formed at the _____ .

j) In the electric arc welding process, the metal _____ also acts as a _____ rod.

## 3. Translate into English:

a) Lötzinn ist eine Metalllegierung, die bei wesentlich niedrigeren Temperaturen schmilzt als die zu lötenden Metalle. Es werden üblicherweise zwei Arten von Lötzinn eingesetzt. Die eine Art ist eine Legierung aus Zinn und Blei, die weicher als die andere Art Lötzinn ist. Die andere Legierung aus Kupfer und Zink wird Silber-Lot genannt.

b) Es ist sehr wichtig, dass die zu verlötenden Metallflächen zunächst gründlich gereinigt werden. Hierzu wird Flussmittel eingesetzt, welches Oxide und Fett von den Metalloberflächen entfernt.

c) Beim Hartlöten muss ein anderes Lot und mehr Wärme eingesetzt werden, um die Schmelztemperatur des Hartlotes von ca. 600°C zu erreichen. Eine hartverlötete Verbindung ist wesentlich stärker als eine normal verlötete.

d) Die Temperatur der Flamme in einem Acetylen-Sauerstoff-Schweißbrenner kann maximal Werte von ungefähr 3000°C erreichen. Der Schweißbrenner erhitzt die Metallteile, die verbunden werden sollen, sowie einen Füllstab, der in die Flamme an die Verbindung gehalten wird.

# 6 The casting of metals

The production of metal objects by the process of casting has been practiced for thousands of years. This is probably the quickest and most economical way of producing a metal component, particularly a large and complex one.

There are many ways in which metals can be cast and the oldest of these is called *sand casting*. In this method, the molten metal is poured into a hollow space in a box filled with sand. The box with its hollow space is called a *mould*, and the hollow space has approximately the *same shape* and *size* as the object which has to be produced as a casting. The manufacture of a casting is usually done in four stages as follows:

1. Making of the pattern

2. Making of the mould

3. Pouring in of the metal

4. Removal of the casting from the mould

The *pattern*, which is a *replica* of the object to be cast is usually made out of wood or metal. Fig 6.1 shows an object which has to be made by the casting process, and Fig 6.3 shows a pattern made out of wood for this object. The wooden pattern is made in *two halves* which can be fitted together by *dowel pins*. Fig 6.4 shows the mould ready for pouring in the molten metal.

The metal is poured into the *mould,* and solidifies as it cools. The casting can finally be taken out by removing the surrounding sand. The pattern and the resulting casting are often made slightly larger than the final form of the object, so that the casting can be machined to *definite dimensions*. This extra allowance in the size of the casting is called *a machining allowance*. The sand casting process is particularly useful when the castings have to be large and sturdy. The bed of a machine tool, or an engine block, are good examples of sand castings.

Another important method of producing castings is called *die casting*. There are two variations of this method - *gravity die casting,* and *pressure die casting*. In both these methods the *moulds are permanent* in character, and are made out of steel. The moulds are very precisely made, and the castings produced are very *accurate* in their *dimensions*, with the result that no machining is required. They also have an excellent *surface finish*. The disadvantage of such a method, is that only *low melting point* metals and alloys, like zinc and aluminium alloys can be used. Such castings are *stronger* than similar (low melting point metal and alloy) castings made by the *sand casting* process (see chap.7, page 38). However they are *not so strong* as castings made of high melting point metals and alloys like *steel* or *brass*.

In *gravity diecasting,* the molten metal is poured into a permanent mould, in a manner similar to the sand casting process. In the *pressure die casting* process, the molten metal is *forced* into *the split dies* of the mould under pressure, with the result that the castings produced are *stronger* and *better finished* than those produced by the gravity die casting process. Typical examples of die castings are the carburettor and the fuel pump of a motor car. The head of a motor car engine is also usually a die casting made of aluminium alloy.

# Vocabulary

| | | | |
|---|---|---|---|
| **casting** | Guss *m* | **molten metal** | flüssiges Metall *n* |
| **carburettor** | Vergaser *m* | **particularly** | besonders *adv* |
| **character** | Eigenart *f* | **pattern** | Muster *n* |
| **complex** | komplex *adj* | **pour** | gießen *v* |
| **dowel pin** | Dübel *m* | **practice** | ausführen *v* |
| **economical** | sparsam *adj* | **pressure diecasting** | Spritzguss *m* |
| **engine head** | Zylinderkopf *m* | **probably** | wahrscheinlich *adj* |
| **force into** | einpressen, hineintreiben *v* | **production** | Herstellung *f* |
| **fuel pump** | Benzinpumpe *f* | **sand casting** | Sandguss *m* |
| **gravity die casting** | Gießen mit Schwerkraft *v* | **shape** | Form *f* |
| **important** | wichtig *adj* | **size** | Größe *f* |
| **machining allowance** | Bearbeitungszugabe *f* | **storage** | Lagerung *f* |
| **method** | Methode *f* | **sturdy** | stark *adj* |
| **die casting** | Spritzguss *m* | | |

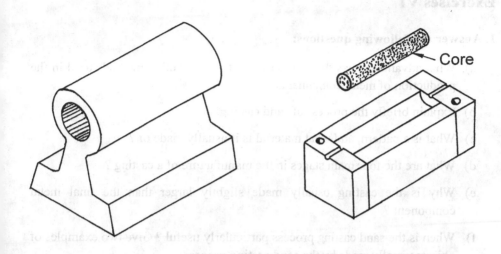

Fig 6.1 Finished casting with a cored       Fig 6.2 Method of producing cores
        hole

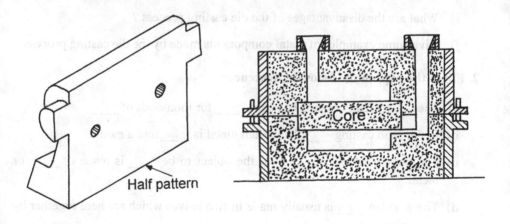

Fig 6.3 One half of a wooden pattern       Fig 6.4 A mould ready for pouring in
                                                   metal

# Exercises VI

### 1. Answer the following questions:

a) What advantages has the method of casting over other methods used in the production of metal components ?

b) Explain briefly the process of sand casting.

c) What is a pattern, and what material is it usually made of ?

d) What are the four main stages in the manufacture of a casting ?

e) Why is the casting usually made slightly larger than the final metal component ?

f) When is the sand casting process particularly useful ? Give two examples of objects usually made by the sand casting process.

g) What are the two main types of die casting processes, and what advantages do die castings have over sand castings ?

h) What kind of material is used for the construction of die casting moulds ?

i) What are the disadvantages of the die casting process ?

j) Give some examples of metal components made by the die casting process.

### 2. Fill in the gaps in the following sentences:

a) The _____ of sand casting has been _____ for thousands of _____ .

b) In the sand casting _____ the molten metal is _____ into a mould.

c) The pattern which is a _____ of the object to be _____ is made of _____ or metal.

d) The wooden _____ is usually made in two halves which are held together by _____ pins.

e) The pattern and the _____ casting are usually made _____ than the final _____ of the finished object.

f) The _____ allowance in the size of the casting is called a _____ allowance.

g) The two methods of producing die castings are called ____ and ____.

h) Die casting moulds are very _____ made, and the castings produced are very accurate in their ____.

i) The disadvantage of the die casting process is that only ____ melting point ____ can be used.

j) In the pressure die casting ____, the molten metal is forced into the dies under ____.

**3. Translate into English:**

a) Der Prozess des Gießens ist wahrscheinlich die schnellste und einfachste Art, einen Metallgegenstand herzustellen, insbesondere bei großen und komplexen Gegenständen.

b) Beim Sandgussverfahren wird das geschmolzene Metall in einen Hohlraum in einem mit Sand gefüllten Behälter gegossen. Der Hohlraum entspricht in Größe und Form in etwa dem Objekt, das gegossen werden soll.

c) Das Muster und der Guss sind normalerweise etwas größer als die fertige Metallkomponente. Dieses Übermaß bezeichnet man als Bearbeitungszugabe.

d) Gussformen, die beim Spritzgussverfahren verwendet werden, sind aus Stahl gefertigt und permanent in ihren Eigenschaften. Die Gussformen sind präzise gefertigt und die hergestellten Gussstücke sehr genau in ihren Abmessungen.

e) Der Nachteil der Spritzgussmethode ist, dass nur niedrigschmelzende Metalle und Legierungen wie Zink und Aluminiumlegierungen in diesem Prozess verwendet werden können. Jedoch haben mit diesem Verfahren hergestellte Gussstücke den Vorteil, dass eine maschinelle Bearbeitung nicht nötig ist.

# 7 The heat treatment of metals

The mechanical properties of metals can be changed to a remarkable extent by subjecting them to heat treatment. Steel is *unique* in its *ability to exist* both as a *soft material* which can be easily machined, and after heat treatment as a *hard material* out of which *metal cutting tools* can be made.

Metals in their usual state consist of a large number of very small *metal crystals interlocked* together, as shown in Fig 7.1. Metals are called *poly-crystalline materials*. The mechanical properties of metals like *hardness, strength,* and *brittleness*, depend on the *size, shape, and orientation* of these *crystals* or *grains*.

### The effect of the rate of cooling and grain size on the strength of a casting

When a casting is made by pouring molten metal into a mould, the *grain structure (or microstructure)* of the casting depends on the rate of cooling of the metal. Slow cooling as in a sand casting results in a *coarse grain structure*, while rapid cooling as in a diecasting results in a *fine grain structure* (Fig 7.1). An object which has a fine grain structure is stronger than an object which has a coarse grain structure. From this it follows that *die castings* are *stronger* than *sand castings*.

### Stress relieving

Metals which undergo plastic deformation by cold rolling, hammering, etc. become *work (or strain) hardened* and are in *a state of stress*. Plastic deformation processes change the grain structure of the metal causing the grains to become elongated (Fig 7.1(c)). The metal becomes brittle and can break easily during use.

A stressed metal component can break easily, and it is important that the stress be removed before it is used. This can be done by the *stress relieving* process in which the steel components are heated to about 600°C for 1 or 2 hours. In this case the grain structure changes very little. This process is *relatively cheap* because of the small amount of energy required, and has the advantage that the metal surface is *not spoilt* by *oxidation* or *scaling*.

If the metal is heated for many hours at this temperature, *recrystallization* takes place and the grain structure reverts to its original state (Fig 7.1 (d)).

## Annealing

The annealing process *softens* steel, *removes stress*, and improves the *toughness*, *ductility,* and *machinability* of the metal. In this process, steel is heated for many hours at a temperature of about 700°C. The furnace is then switched-off, allowing both the components and the furnace to cool slowly together.

## Normalizing

In the normalizing process, steel is heated to a temperature slightly above the *upper critical* temperature line (Fig 7.2) for a short period of time, and allowed to cool in air at room temperature. The steel now acquires a completely new *fine grain structure* which makes it *stronger,* more *homogeneous,* and *free of stress.*

## Hardening

Hardened steel is used to make components which *resist wear*, and also to make *metal cutting tools.* Steel is hardened by heating it to a suitable temperature above the *upper critical line* (Fig 7.3) and then *quenching* in oil or water. The steel acquires a needle like microstructure called *martensite* which makes it glass hard.

## Tempering

Steel that has been quenched is too *brittle* for *normal use* and has to be tempered. Tempering is a process which *reduces* the *brittleness* and *hardness* of the steel, while **increasing** its *toughness*. Here the steel is heated to an appropriate temperature below the *lower critical* temperature line (Fig 7.3 ) and allowed to cool slowly.

The temperatures used for tempering depend on the purpose for which the component is used. If the surface is polished, the colour of the *oxide film* formed gives an indication of the temperatures reached. Cutting tools, razors, etc. which need to be *hard* but *not so tough*, are tempered at 200-250°C and acquire a *pale yellow colour*. Springs, screwdrivers, etc which need to be *tough* but *not so hard,* are tempered at a higher temperature of 300-350°C and acquire a *deep blue colour*.

## Case hardening

Sometimes it is necessary to have a mild steel component which has a hard surface and a tough interior. Such a component can be produced by *case hardening* which is carried out in two stages as explained below.

a) **Carburization** - In the first stage of the case hardening process, **mild steel components** are packed in boxes with **carbon rich** substances and heated for several hours at temperatures of 900-950°C. Carbon is absorbed into the surface converting the mild steel into a **high carbon steel** up to a depth of about 1 mm.

b) **Refining the core and hardening the surface** - Prolonged heating during carburization gives the core a coarse grain structure. The grain structure of the core can be **refined** (Fig 7.1(e)) and the **component made tougher**, by heating to about 770°C and quenching in water. The surface will now be extremely brittle, and this **brittleness** has **to be reduced** by **tempering** at a suitable temperature.

## Other surface hardening methods

a) In **induction hardening,** high frequency currents are induced on the surface of a steel component by a coil which carries a high frequency electric current. The heat generated causes the surface temperature to rise to the hardening temperature. The component is **immediately quenched** by moving it through a spray of water.

b) In **flame hardening**, the surface of the steel component is heated by a flame to the hardening temperature, and then **quenched** by a **spray of water**.

c) In **nitriding**, an alloy steel component is maintained for many hours in ammonia gas at about 550°C. The surface of the component becomes hard due to the **formation** of **hard nitrides** (of alloying elements like chromium or vanadium) on the surface.

## Vocabulary

| anneal | weichglühen v | interlock | ineinandergreifen v |
|---|---|---|---|
| carburize | aufkohlen v | normalize | normalglühen v |
| case harden | einsatzhärten v | polycrystalline | vielkristallin adj |
| coarse grain | grobkörnig adj | quench | abschrecken v |
| elongate | verlängern, strecken v | refine | rückfeinen, vergüten v |
| fine grain | feinkörnig adj | stress relieve | spannungsarm glühen v |
| grain structure | Gefüge n | temper | anlassen v |

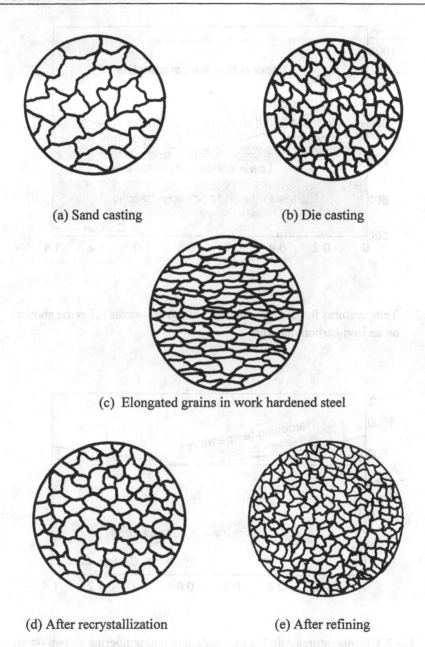

(a) Sand casting                                    (b) Die casting

(c)  Elongated grains in work hardened steel

(d) After recrystallization                    (e) After refining

Fig 7.1 Various grain structures in metals and castings

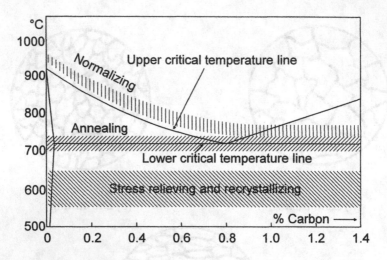

Fig 7.2  Temperatures for normalizing, annealing, and stress relieving shown
           on an iron-carbon diagram

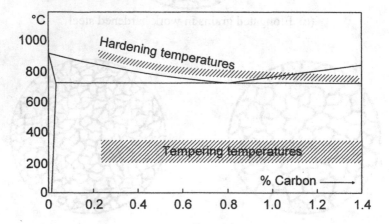

Fig 7.3 Temperatures suitable for hardening and tempering shown on an
           iron-carbon diagram

# Exercises VII

## 1. Answer the following questions:

a) What makes steel such a unique material ?

b) What type of crystal structure do polycrystalline materials have ?

c) How does the rate of cooling affect the grain structure of a casting ?

d) What happens to the grain structure of a metal when it is plastically deformed ?

e) What advantages does the stress relieving process have over other similar heat treatment processes ?

f) How are the mechanical properties of steel changed by the annealing process ?

g) What types of steel components need to be hardened?

h) Why must hardened steel components be tempered ?

i) Why is it necessary to refine a component after it has been case hardened ?

j) What causes the surface of a nitrided component to become hard ?

## 2. Fill in the gaps in the following sentences:

a) Slow cooling as in a ____ results in a ____ grain structure.

b) Metals which undergo ____ become work ____.

c) The ____ process is relatively cheap, and has the advantage that it does not cause the metal surface to become ____.

d) The annealing process ____ steel and ____ its toughness.

e) Hardened steel is used to make ____ which ____ wear.

f) Steel is hardened by heating it to a temperature above the ____ and then ____ in water.

g) Tempering reduces the ____ and hardness of steel while increasing its____.

h) Case hardening produces a component with a ____ and a tough ____.

i) Hardened steel acquires a needle like grain structure called ____ which makes it ____.

j) Carbon is absorbed into the ____ converting it into a ____ steel.

## 3. Translate into English:

a) Das Wärmebehandlungsverfahren verbessert die mechanischen Eigenschaften eines Metalls, wodurch der Fertigungsprozess erleichtert wird. Außerdem widersteht eine wärmebehandelte Metallkomponente erschwerten Einsatzbedingungen.

b) Die mechanischen Eigenschaften eines Metalls wie Härte, Festigkeit und Sprödigkeit hängen von der Größe, Form und Ausrichtung der Mikrokristalle ab.

c) Wenn ein Guss angefertigt wird, hängt die Mikrokristallstruktur des Gussstücks von der Abkühlgeschwindigkeit während des Verfestigens ab. Ist das Abkühlen langsam, wie bei einem Sandguss, werden die Mikrokristalle groß. Kühlt es dagegen schnell ab, so werden die Mikrokristalle erheblich kleiner.

d) Metall wird während des Fertigungsgangs oft starker Spannung ausgesetzt. Diese Spannung muss vor Einsatz der Metallkomponente entfernt werden. Dieses kann dadurch erreicht werden, dass man das Werkstück auf eine Temperatur gerade über 500°C erhitzt und anschließend auskühlen lässt.

# 8 The forging of metals

We have seen that metal components can be produced cheaply and efficiently by the casting process. Cast components however, are usually brittle and break easily. Some metal components like crankshafts, connecting rods, spanners, etc. have to stand up to conditions of *severe stress* when in use. Components produced by the *forging process*, are much stronger than those produced by casting or by machining from solid bar material. The reason for this lies in the fact that the *grain structure* is different. Each metal grain is a minute crystal, and the strength of a metal component depends on the *orientation of the grains*. Fig 8.1 shows the difference in grain structure between a gear tooth that has been machined on the right of the diagram, and a gear tooth that has been forged on the left.

The process of forging is also *more economical* in the use of material, because the metal is forced into the shape of the final component, and no material is wasted as in the machining process.

It involves the *forcing* of *red hot metal* into the shape required. Not all metals can be forged. Cast iron when heated to red heat becomes very *brittle*, and breaks easily when struck with a hammer. Mild steels and many other metals become *ductile* at high temperatures, and can then be hammered or forced into the required shape. In general, there are three ways in which metal components can be produced by forging:

## Hand and hammer forging

Forged components in small quantities were produced by blacksmiths over the ages. A blacksmith was a highly skilled person, who used hand tools to shape a piece of *red hot metal* into the required form. After the beginning of the industrial revolution, large *mechanical hammers* were used. These had the necessary power to make the forging of much larger components possible.

## Drop forging

*Drop forging* is particularly useful, where the production of a large number of medium-size forgings is required. The method involved, uses dies similar to those used in die casting. Two half dies are used, the two halves being kept in close alignment with each other. The downward force of the top die forces the red hot *metal billet* into the *cavity* between the dies. This process is called *drop* or *closed die* forging . Fig 8.2 shows an example of this process.

## Upset forging

Often, several stages of forging are required to produce a metal component. Several stages in the production of a *socket spanner* by the process of *upset forging* are shown in Fig 8.3. In this process, only a part of the component is made red hot. A change in the *cross-sectional area* of the *red hot portion* of the metal component is made by the application of suitable forces.

# Vocabulary

| | | | |
|---|---|---|---|
| alignment | Ausrichtung *f* | ductile | dehnbar, biegbar *adj* |
| billet | Knüppel *m* | drop forging | Gesenkschmieden *n* |
| blacksmith | Schmied *m* | efficient | leistungsfähig *adj* |
| brittle | spröde *adj* | example | Beispiel *n* |
| cheap | billig *adj* | severe | streng *adj* |
| compound | Verbindung *f* | stage | Phase *f* |
| connecting rod | Pleuelstange *f* | stress | Spannung *f*, |
| | | | Beanspruchung *f* |
| crankshaft | Kurbelwelle *f* | suitable | geeignet *adj* |
| crystal grain | Mikrokristall *n* | | |

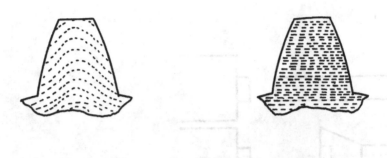

Forged                              Machined from bar

Fig 8.1 Difference in grain structure betweeen a forged and a machined gear tooth

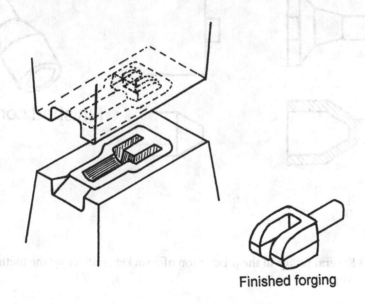

Finished forging

Fig 8.2 Drop forging produced by using split dies

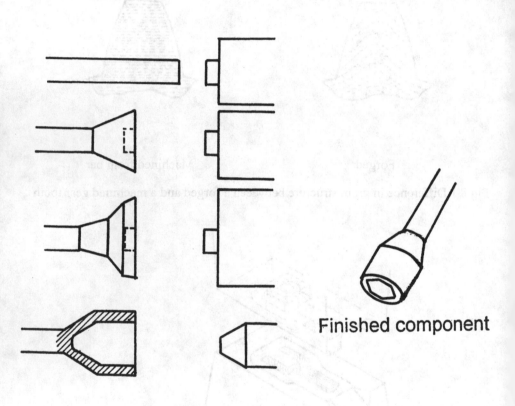

**Finished component**

Fig 8.3 Several stages in the production of a socket spanner by the method of upset forging

# Exercises VIII

## 1.Answer the following questions:

a) What is probably the cheapest way of producing complex metal components?

b) What advantages do forged components have over cast components ?

c) Why is no material wasted in the forging process ?

d) Why cannot cast iron be forged ?

e) How does a blacksmith produce forged components ?

f) When were mechanical hammers first used for forging, and what advantages do they have ?

g) State three ways of producing a forged component.

h) When is drop forging particularly useful ?

i) How is a red hot metal billet transformed into a forging by the drop forging process ?

j) What kind of change in form takes place in the upset forging process ?

## 2.Fill in the gaps in the following sentences:

a) Metal components can be ____ cheaply and ____ by the casting process.

b) Cast components are usually ____ and break ____.

c) Forged components are ____ than cast components or those produced by ____.

d) Forging is more ____ in the use of material, because the metal is ____ into the required shape.

e) Mild steel becomes ____ at high ____, and can be forced into the ____ shape.

f) Forged ____ were produced in small ____ by blacksmiths.

g) Large mechanical ____ were used to ____ large ____.

h) Two ____ dies are used in the drop forging process, and the red hot billet is ____ into them.

i) Often,____ stages of forging are ____ to ____ a forged component.

j) In upset forging, the ____ area of the component is ____ .

## 3. Translate into English:

a) Metallwerkstücke können durch Gießen preiswerter als durch Schmieden hergestellt werden. Mit dem Gussverfahren hergestellte Komponenten sind jedoch sehr spröde und brechen leicht.

b) Das Verfahren des Schmiedens ist sehr ökonomisch im Materialverbrauch, weil das Metall in die Form des endgültigen Produkts gepresst wird und so kein Metall verschwendet wird.

c) Stahl und viele Metalle werden bei hohen Temperaturen dehnbar und können dann in die gewünschte Form gepresst oder gehämmert werden. Gusseisen kann jedoch nicht auf diese Weise geschmiedet werden.

d) Die Methode des Gesenkschmiedens ist besonders nützlich, wenn eine große Anzahl einer Komponente hergestellt werden soll.

e) Beim Verfahren des Gesenkschmiedens werden zwei Halbformen benutzt. Der von der oberen Form ausgeübte Druck presst den rotglühenden Metallknüppel in die Aushöhlung zwischen den Formen.

# 9  Hot and cold bulk deformation processes

Metals are initially produced by the refining of ores or scrap metal in furnaces. The molten metals or alloys are usually cast into *ingots, slabs,* or *billets*. Before metal goods can be manufactured, it is necessary to transform these cast metal forms into *intermediate (or semifabricated) products* like metal sheets, wire, and rods, which are the starting point for the manufacture of more complex products. *Cold* or *hot bulk deformation* processes like *rolling, drawing, extruding,* and *forging* which involve large scale plastic deformation are used for this purpose. Unlike sheet metal working where the changes in thickness are small, bulk deformation processes cause *large changes* in *thickness, diameter,* or *other dimensions* of the cast metal forms.

### Strain hardening caused by bulk deformation processes

Stress is required to plastically deform a metal, and this causes *strain (or work) hardening* (see chapter 2). The metal becomes brittle, and *heat treatment* may be needed to soften the metal and bring it into a more normal state (see chap 7).

### Hot working processes

In hot working processes, the metal is heated to a temperature which is usually above the *recrystallization temperature*, before it is subjected to plastic deformation. The advantages and disadvantages of hot working are as follows.

- The metal flows more easily at high temperatures, and consequently smaller forces and less power are required for plastic deformation.

- Ductility is high, and hence large deformations without fracture are possible. Complex shapes can be generated without much difficulty.

- Work hardening does not take place, and the components produced (especially forged components) are extremely strong.

Disadvantages are that *dimensional tolerances* are *low,* and that the metal surface is *spoilt* by *oxidation* and the *formation of scale*. In addition, *large amounts of energy* are required to heat the object before it is deformed.

## Cold working processes

Cold working usually means working at *room temperature*. The advantages of cold working are considerable.

- Better surface finish, closer tolerances, and thinner walls in products can be realized.

- The increase in strength due to strain hardening can be *retained* if required, or if preferred the metal can be returned to a *ductile state* by heat treatment.

Disadvantages are the high flow stresses involved, which make high tool pressures and large power requirements necessary.

## Rolling

Rolling is the most important of the bulk deformation processes. In *flat rolling*, the thickness of a slab is reduced to produce a thinner and longer but only slightly wider product (Fig.9.1). Initially, cast slabs are rolled by *hot rolling processes*. Hot rolled plates or sheets have *rough surfaces* and *poor dimensional tolerances*. They are *relatively thick*, and are used in applications like ship building, boiler making, and for the manufacture of welded machine frames.

Thinner sheets are manufactured from the hot rolled sheets by *cold rolling*. Cold rolled sheets have a *better surface finish* and *tighter tolerances*. In addition to flat rolling, hot or cold *shape rolling* can be used to produce long bars, rods, etc. each having a *uniform* but *different cross-section* (Fig.9.2).

In *ring rolling* (Fig.9.3), and *tube rolling* (Fig.9.4), pierced billets and centering mandrels are used to manufacture hollow products. *Thread rolling* is used to manufacture *screws, taps,* etc. which have stronger screw threads than those produced by screw cutting processes.

## Drawing

In the drawing process, the material is pulled through a die of gradually *decreasing cross-section*. Most types of wire having circular, square, or other types of cross-section, are manufactured by drawing (Fig.9.5). Metal wire is the starting point for the manufacture of a number of products like *screws, nails, bolts,* and *wire frame structures*. *Seamless tubes* can also be produced by a drawing process (Fig.9.6).

Larger tubes are usually hot drawn, but smaller tubes below a certain diameter have to be cold drawn.

**Extrusion**

Long rods, tubes, etc. of uniform cross-section can be produced by extrusion. In this process, the material is under pressure and is *forced to flow through a die.* The cross-sections of the extruded products can have different sizes and shapes, depending on the size and shape of the opening in the die.

- In *forward or direct extrusion*, the extruded product moves in the same direction as the punch which pushes the material (Fig 9.7).

- In *reverse, indirect (or back) extrusion*, the product moves in the opposite direction to the direction of movement of the punch (Fig.9.8).

- *Hot and cold extrusion* are both possible, and *hollow products* can be extruded by using a centered mandrel (Fig.9.9).

# Vocabulary

| | | | |
|---|---|---|---|
| amount | Menge *f* | oxidation | Oxidierung *f* |
| bar | Stange *f*, Stab *m* | plate | Platte *f* |
| billet | Knüppel *m* | power | Leistung *f* |
| bulk deform | große Masse umformen *v* | realize | verwirklichen *v* |
| disadvantage | Nachteil *m* | refine | raffinieren *v* |
| extrude | strangpressen *v* | rod | Rundstab *m* |
| flow | fließen *v* | scale | Kesselstein *m*, Kruste *f* |
| force | Kraft *f* | screw thread | Gewinde *n* |
| forge | schmieden *v* | seamless | nahtlos *adj* |
| generate | erzeugen *v* | semifabricated | halbfertig *adj* |
| heat treatment | Wärmebehandlung *f* | slab | Platte *f* |
| ingot | Gussblock *m* | soften | weich machen *v* |
| mandrel | Dorn *m* | tight | eng *adj* |
| ore | Erz *n* | transform | umwandeln *v* |

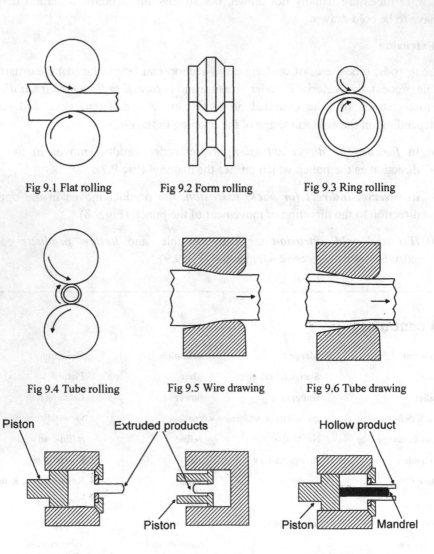

Fig 9.1 Flat rolling          Fig 9.2 Form rolling          Fig 9.3 Ring rolling

Fig 9.4 Tube rolling          Fig 9.5 Wire drawing          Fig 9.6 Tube drawing

Piston                        Extruded products             Hollow product

Fig 9.7 Forward              Fig 9.8 Reverse               Fig 9.9 Extrusion
    extrusion                    extrusion                  of hollow products

Piston                        Piston                        Mandrel

# Exercises IX

## 1. Answer the following questions:

a) How are metals initially produced ?

b) Why is it necessary to tranform cast ingots into intermediate products ?

c) What is strain hardening ?

d) How can a strain hardened metal be brought back into a normal state ?

e) Why are the forces required to cause plastic deformation in a hot working process smaller than those required in a cold working process ?

f) What advantage does the thread rolling process have over the thread cutting process ?

g) How are thin metal sheets produced ?

h) Name some types of products that are manufactured from metal wire.

i) What are the most important characteristics of hot rolled plates ?

j) In what way are products manufactured by cold working processes better than those manufactured by hot working processes ?

## 2. Fill in the gaps in the following sentences:

a) Metals are initially produced by the ____ of ores and scrap metal in ____.

b) Stress is required to plastically ____ a metal, and this causes ____ .

c) The disadvantages of hot working are that ____ tolerances are low, and that the surface is spoilt by ____.

d) Ductility is high, and large ____ without ____ are possible.

e) The increase in ____ due to ____ hardening may be retained if required.

f) In flat rolling, the ____ of a slab is ____ to produce a longer product.

g) Hot rolled sheets have ____ surfaces, and ____ tolerances.

h) In ring rolling, pierced ____ and centering ____ are used to produce hollow products.

i)  Smaller tubes below a certain ____ have to be cold ____ .

j)  Rolling is the most ____ of ____ processes.

## 3. Translate into English:

a) Bevor Metallwaren hergestellt werden können ist es notwendig, die gegossenen Metallformen in Zwischenformen wie Bleche, Draht und Rundstäbe zu überführen, welche Ausgangspunkt für die Herstellung komplexerer Produkte sind.

b) Anders als bei Blecharbeiten, wo Änderungen der Stärke gering sind, verursacht das Umformen großer Massen große Änderungen in Stärke, Durchmesser oder anderer Größenparameter der gegossenen Metallformen.

c) Beim Prozess des Ziehens wird das Material durch ein Schneideisen mit allmählich abnehmendem Querschnitt gezogen. Draht ist der Ausgangspunkt für die Herstellung einer Vielzahl von Produkten wie Schrauben, Nägel, Muttern und Drahtrahmen.

d) Lange Stäbe, Röhren, usw. einheitlichen Querschnitts können durch Strangpressen hergestellt werden. Bei diesem Prozess ist das Material unter Druck und wird gezwungen durch ein Schneideisen zu fließen. Die Querschnitte stranggepresster Produkte haben unterschiedliche Größen und Formen, abhängig von der Größe und Form des Schneideisens.

# 10  The transmission of power

A working machine is composed of many components, some of which are static and some of which are moving. There are many kinds of motion, of which *rotatory motion*, *linear motion*, and *reciprocating motion* are among the most important.

In *rotatory (or rotary) motion*, the object is rotating around an axis or a point in a *clockwise* or an *anticlockwise* sense. In *linear motion* the object moves in a straight line. In *reciprocating motion* the object moves along a straight line in one direction, and then back again along the same straight line to its starting point. A good example of reciprocating motion is the motion of a piston in the cylinder of an automobile engine.

A rotating component which is found in nearly all machines is a *shaft*. A shaft is a cylindrical rod usually made of steel on which various other components like gear wheels, pulleys, etc. are often mounted. The *supports* in which a shaft rotates are called *bearings*.

Two types of bearings are commonly used. The first type is the *plain bearing*, and the other type consists of *ball* and *roller bearings*. A plain metal bearing is made of a metal or an alloy which is usually softer than steel. *Bronze* or *white metal* are suitable alloys for plain bearings. Quite often a plain bearing is made in two halves and is called a split bearing. Fig 10.1 shows *split shell bearings* fitted on the *big end* of an automobile connecting rod. On the *small end* of the connecting rod there is a one piece hollow cylindrical bearing called a *gudgeon pin* (or bush). Plain bearings need to have a thin film of oil between the contact surfaces. The maintenance of such a film is called *lubrication*.

*Ball bearings* have less friction than plain bearings, and they are also more easily replaced when renewal becomes necessary. A ball bearing consists of hardened steel spheres running between two precision ground hard cylindrical races as shown in Fig 10.2. The components of a taper *roller bearing* are shown in Fig 10.3. The rollers are located in a cage between an outer cup and an inner cone. Lubrication of ball or roller bearings is simpler than the lubrication of plain bearings. They only need an occasional application of *grease*.

## Direct coupling of two shafts which are in the same line (in-line shafts)

A direct coupling is often used to make a semi-permanent connection between two shafts which are in the same straight line. *Rigid couplings* can only be used if the shafts are in perfect alignment. *Flexible couplings* made of rubber or metal springs may be used when slight *misalignment* between the shafts exists. *Universal joints* are used to couple shafts which have larger values of misalignment.

## Coupling between two rotating shafts

A rotating shaft has *mechanical energy* and quite often it is necessary to transfer this energy to another shaft which may be rotating at a different speed. The transfer of energy can be accomplished by having some form of *coupling* between the shafts. Some of the ways in which this may be done are as follows:

1. by using a belt and two pulleys (Fig 10.4)

2. by using gear wheels (Fig 10.5)

3. by using a chain and two sprockets (also called sprocket wheels) (Fig 10.6)

## Clutches

Clutches are special couplings which allow two coupled shafts to be *disengaged* (or separated from each other), even when they are *rotating*. They are used for example in the *gear boxes* of machine tool drives, car engines, etc.

## Conversion of one type of motion into another

It is often necessary to change one type of motion into another with a minimum loss of energy. A good example is the case of the conversion of the *reciprocating motion* of a piston into the *rotational motion* of a crankshaft. Fig 10.7 shows how this is accomplished in a car engine. The essential device required here is the *crank*, which makes the *conversion* possible. Sometimes it is necessary to convert a rotational motion into a linear motion. Fig 10.8 shows how this can be accomplished by using a *rack and pinion*.

# Vocabulary

| | | | |
|---|---|---|---|
| **accomplish** | ausführen *v* | **grease** | Fett *n* |
| **alignment** | Anordnung *f* | **half (pl. halves)** | Hälfte *f* |
| **amongst** | unter *pr* | **important** | wichtig *adj* |
| **anticlockwise** | entgegen dem Uhrzeigersinn | **linear** | linear *adj* |
| **axis** | Achse *f* | **lubrication** | Schmieren *n* |
| **bearing** | Lager *n* | **machine tool** | Werkzeugmaschine *f* |
| **belt** | Riemen *m* | **minimum** | Minimum *n* |
| **clockwise** | im Uhrzeigersinn | **motion** | Bewegung *f* |
| **clutch** | schaltbare Kupplung *f* | **necessary** | notwendig *adj* |
| **conversion** | Umsetzung *f* | **often** | oft *adv* |
| **coupling** | Kupplung *f* | **pinion** | Ritzel *n* |
| **crank** | Kurbel *f* | **power** | Kraft *f* |
| **crankshaft** | Kurbelwelle *f* | **rack** | Zahnstange *f* |
| **direction** | Richtung *f* | **reciprocating motion** | pendelnde Bewegung *f* |
| **disengage** | ausschalten *v* | **rigid coupling** | starre Kupplung *f* |
| **energy** | Energie *f* | **rotatory, rotary** | drehbeweglich *adj* |
| **engine** | Motor *m* | **spring** | Feder *f* |
| **essential** | wesentlich *adj* | **torque** | Drehmoment *n* |
| **friction** | Reibung *f* | **universal joint** | Gelenkkupplung *f* |
| **gear wheel** | Zahnrad *n* | | |

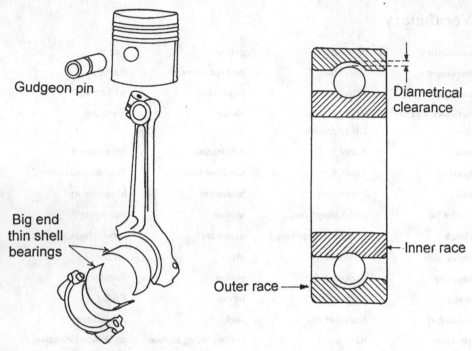

Gudgeon pin

Big end
thin shell
bearings

Diametrical
clearance

Inner race

Outer race

Fig 10.1 Split shell bearings and
gudgeon pin

Fig 10.2 Cylindrical races in a ball
bearing

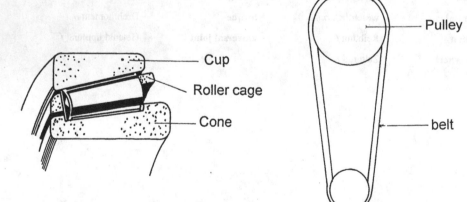

Cup

Roller cage

Cone

Pulley

belt

Fig 10.3 Components of a tapered
roller bearing

Fig 10.4 Belt and  pulley drive

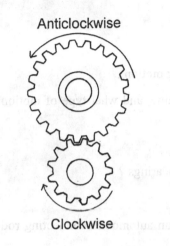

Fig 10.5 Gear wheel drive

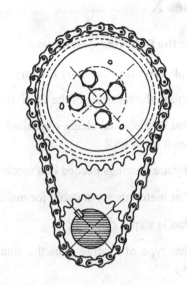

Fig 10.6 Chain drive with sprocket
wheels

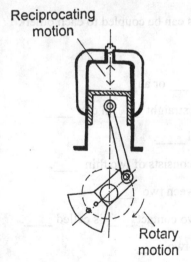

Fig 10.7 Conversion of reciprocating
motion to rotary motion

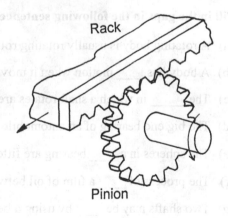

Fig 10.8 Rack and pinion drive

# Exercises X

### 1. Answer the following questions:

a) Explain what is meant by rotatory motion .

b) Explain the meaning of the term  reciprocating motion .

c) What is the usual shape of a shaft in a machine, and what type of motion does it undergo ?

d) Why are bearings needed in a machine ?

e) What materials are suitable for making plain bearings ?

f) What is a split bearing ?

g) What type of bearing does the small end of an automobile connecting rod have ?

h) Describe the construction of an ordinary ball bearing .

i) What is the purpose of lubrication ?

j) State three ways in which two rotating shafts can be coupled to each other .

### 2. Fill in the gaps in the following sentences:

a) A rotating body is usually rotating round a ____ or an ____ .

b) A body has ____ motion when it moves in a straight ____ in one ____ .

c) The ____ in which a shaft rotates are called ____ .

d) The big end bearing of an automobile ____ consists of two thin ____ .

e) The spheres in a ____ bearing are fitted between two ____ .

f) The process of ____ a film of oil between two contact ____ is called ____ .

g) Two shafts may be ____ by using a belt and two ____ .

h) The ____ bearing on an automobile connecting rod is called a ____ .

i)  The rollers in a roller bearing are ____ between an ____ cup and an inner ____.

j)  Two ____ may be coupled using a chain and two ____ wheels.

## 3. Translate into English:

a)  Führt ein Körper eine Rotationsbewegung aus, so dreht er sich üblicherweise gegen oder im Uhrzeigersinn um eine Achse. Ist die Bewegung
gleichförmig, so bewegt sich der Körper auf einer geraden Linie.

b)  Eine Maschine ist aus vielen Teilen zusammengesetzt von denen einige sich
bewegen. Eine Welle ist ein zylindrischer Stab, der üblicherweise aus Stahl
hergestellt wird und in Lagern, die sein Gewicht abstützen, rotieren kann.

c)  Lager treten gewöhnlich in Form von Gleitlagern, Kugellagern oder
Rollenlagern auf. Für Gleitlager werden besondere Legierungen verwendet,
z.B. Bronze.

d)  Kugellager setzen sich aus Stahlkugeln und zwei Lagerringen, zwischen
denen die Kugeln gehalten werden, zusammen. Die Schmierung solcher
Lager ist einfach, da sie nur gelegentlich etwas Schmierfett benötigen.

e)  Eine rotierende Welle besitzt mechanische Energie. Es ist oft nötig, diese
Energie einer anderen Welle zuzuführen. Dieses kann durch zwei Zahnräder
oder zwei Rollen und einem Gurt erfolgen.

# 11 Gears and gearing

Gears (or gear wheels) are *toothed wheels* which are usually used to couple two or more rotating shafts. The shafts may be rotating at the *same speed* or at *different speeds*, and can have axes which are pointing in the *same direction* or in *different directions*. The *ratio* of the *speeds of rotation* of two shafts coupled by gear wheels, depends on the *ratio* of the *number of teeth* in the wheels.

## Spur gears

There are many types of gears of which the *spur gear* is the most common. A spur gear is a wheel with teeth cut parallel to its axis of rotation. These gears are used to couple shafts that are *parallel* to each other. The larger wheel is called the *gear* and the smaller wheel is called the *pinion*. Fig 11.1 shows two spur gears, and some of the dimensions (like pitch diameter and working depth) required for their design. In addition to spur gears with *external teeth* (Fig 11.1) it is also possible to have gears with *internal teeth* (Fig 11.2).

## Bevel gears

In a *bevel gear,* the teeth are cut in such a way that they lie on a *conical surface* and appear to meet at the apex of a cone. Fig 11.3 shows a *bevel gear* and a *pinion*, and also the cones corresponding to each gear. Bevel gears are used to couple shafts which are *at an angle* to each other (usually 90 degrees).

## Helical gears

The teeth of a *helical gear* lie on a *cylinder* and are cut at an angle to the axis of rotation of the gear. Helical gears can be used to transmit motion between two parallel shafts, or between shafts at an angle to each other (Fig 11.4). Helical gears are *quieter* and *smoother* in operation than spur gears. However, the friction and the resulting heat and wear produced, are greater than for spur gears. Helical gears are usually placed in an *oil bath* (as for example in an automobile gear box), so that the wear on the teeth will be minimized.

## Rack and Pinion

A rack and pinion is a device in which a spur gear wheel called the *pinion*, meshes with a flat piece of metal on which gear teeth have been cut called the *rack*. When the pinion rotates, the rack moves in a straight line.This device converts rotatory motion into linear motion (Fig 10.8). A good example of the use of a rack and pinion is the drill feed mechanism of a drilling machine.

## Worm and worm gear

Worm gears are usually used for heavy duty work  where a *large ratio of speeds* is required. They are extensively used as *speed reducers*. The *worm* is the driving wheel, and is a cylinder with teeth cut in the form of a *screw thread*. The *worm gear* is the driven wheel, and has teeth cut at an angle to the axis of rotation. The axes of the wheels are usually at right angles to each other  (Fig 11.5).

## Gear boxes

The *ratio* of the *speeds of rotation* of two shafts, is dependent on the *ratio* of the *number of teeth* in the gears. It is often necessary to *change* this gear ratio *repeatedly*. This is normally done in several stages, through several pairs of gears and shafts. A device for doing this smoothly and rapidly is called a *gear box*. Gear boxes are commonly used in *machine tools* and in *automobiles*. Fig 11.6 shows the gear box of a lathe.

## The manufacture of gears by gear cutting and gear generating processes

*Gear cutting* can be done on a milling machine fitted with special milling cutters.The edges of these cutters have the profile of the *space between adjacent gear teeth*. Since only one tooth can be cut at a time, this method is unsuitable for mass production. Mass production of gear wheels with *mathematically accurate* tooth profiles can be achieved by using *gear generating processes*. Two of these processes are briefly described below.

The *gear shaping process* uses a cutting tool in the form of a *gear (pinion) wheel.* The cutting tool and the gear blank (which has to be cut), are made to revolve at speeds corresponding to their being coupled with each other. The cutting action is produced by *reciprocating the cutting tool* in a direction parallel to the axis of the blank. Spur and helical gears (both internal and external) can be generated by this process. A *rack* can also be used as the cutting tool instead of a pinion.

*The gear hobbing process* uses a cutting tool called a *hob* which has the shape of a *worm*. Slots are cut in the worm so that it acquires *cutting teeth*. During the cutting process, the hob and the gear blank are made to revolve at appropriate speeds corresponding to their being coupled with each other. The motion is similar to that of a *worm and worm gear* (see Fig 11.5). By *tilting the axis* of the hob through appropriate angles, both spur and helical gears can be generated. This process cannot be used to generate internal gears.

**Other methods for the production of gears**

Many other methods are used for the production of gears. Among these are blanking, broaching, casting, die casting, cold extruding, cold pressing, forging, grinding, rolling, and sintering.

# Vocabulary

| | | | |
|---|---|---|---|
| **bevel gear** | Kegelrad *n* | **hobbing process** | wälzfräsen *v* |
| **cone** | Kegel *m* | **internal** | innerlich *adj* |
| **depend** | abhängen *v* | **maintain** | erhalten *v* |
| **depth** | Tiefe *f* | **mechanism** | Mechanismus *m* |
| **design** | entwerfen *v* | **quiet** | ruhig *adj* |
| **determine** | bestimmen, entscheiden *v* | **rack and pinion** | Zahnstangengetriebe *n* |
| **drive** | treiben *v* | **rapid** | schnell *adj* |
| **extensive** | umfassend *adj* | **ratio** | Verhältnis *n* |
| **external** | äußerlich *adj* | **reduce** | vermindern *v* |
| **gear forming** | wälzstoßen *v* | **spur gear** | Stirnrad *n* |
| **helical gears** | Zahnräder mit Schrägverzahnung *n* | **worm gear** | Schneckengetriebe *n* |

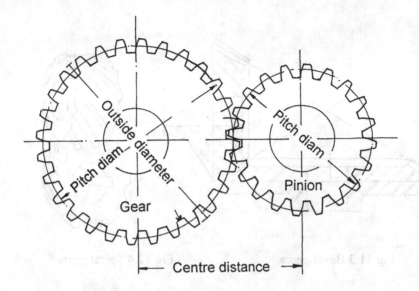

Fig 11.1 Spur gears

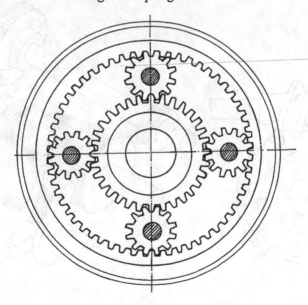

Fig 11.2 Internal epicyclic gearing

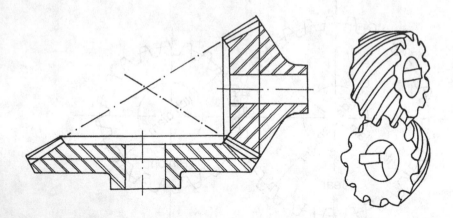

Fig 11.3 Bevel gears                    Fig 11.4 Spiral gears

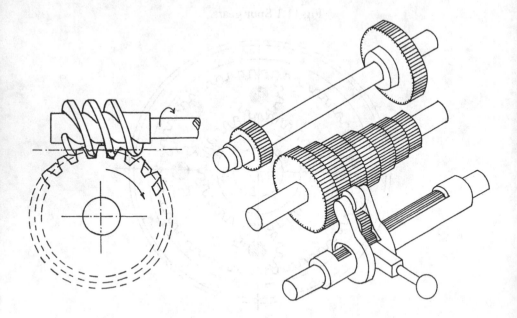

Fig 11.5 Worm and worm gear        Fig 11.6 Gear box fitted to a lathe

# Exercises XI

## 1. Answer the following questions:

a) For what purpose are gear wheels used in a machine ?

b) Explain how the ratio of the speeds between shafts, or the direction of the axes of rotation of two shafts coupled by gears can be changed.

c) What are spur gears, and for what purpose are they used ?

d) What kind of teeth do bevel gears have ?

e) What kind of teeth do helical gears have ?

f) Compare the advantages and disadvantages of using spur and bevel gears.

g) Why are helical gears usually placed in an oil bath ?

h) For what purpose can a rack and pinion be used ?

i) For what type of work are worm gears used ?

j) Why are gearboxes needed in cars and machine tools ?

## 2. Fill in the gaps in the following sentences:

a) Gear wheels are used ____ two or more ____.

b) A spur gear has ____ cut parallel to its axis of ____.

c) The larger gear wheel is ____ the gear, and the smaller gear wheel ____.

d) The teeth in a bevel gear lie on a ____ surface and appear to ____ at the apex of a ____.

e) The teeth in helical gears lie on a ____, and are cut at an angle to the axis of ____.

f) Helical gears are usually ____ in an ____ in order to minimize the wear on the teeth.

g) A rack and pinion ____ rotatory motion into ____ motion.

h) Worm gears are used for ____ work where a large ratio of ____ is required.

i) Helical gears are quieter and ____ in operation than ____.

j) When two gears are ____ together, the ratio of the speeds of rotation of the ____ depends on the ratio of the ____ in the gears.

## 3. Translate into English :

a) Ein Zahnrad überträgt im Verbund mit anderen Zahnrädern Bewegung von einem Teil eines Mechanismus zum anderen.

b) Bei schrägverzahnten Zahnrädern ist die Reibung und die resultierende Wärme und Abnutzung höher als bei anderen Getrieben. Um die Abnutzung zu vermindern, werden diese Getriebe häufig in ein Ölbad gesetzt.

c) Bei einem Kegelrad sind die Zähne so geschnitten, dass sie auf einer kegelförmigen Oberfläche liegen und sich im Scheitel des Kegels zu treffen scheinen.

d) Die Zähne eines Schrägzahnrades sind auf einem Zylinder eingeschnitten und bilden einen Winkel mit der Drehachse des Getriebes. Schrägverzahnte Zahnräder sind leiser und laufen ruhiger als Stirnradgetriebe.

e) Wenn zwei an Wellen befestigte Zahnräder ineinandergreifen, hängt das Verhältnis der Drehgeschwindigkeiten der Wellen von dem Verhältnis der Anzahl von Zähnen der beiden Zahnräder ab.

## 12   Screws, keys, splines, and cotters

### Screws and screw threads

*Screws* and *bolts* are commonly used for fastening various types of components. A screw has a *screw thread* which is a ridge in the form of a spiral on the surface of a cylinder or a cone. There are many shapes and types of screw threads, each being made to a *definite specification*. Some of the terms used in defining such a specification are given below.

- *The angle of a thread* is the angle between the two inclined faces of the thread (Fig 12.1).

- *The pitch* is the distance between a point on a screw thread and the corresponding point on the next thread, measured parallel to the screw's axis.

- *The lead* is the distance advanced by the screw when it is rotated through 360°. In a *single thread* screw, the lead and pitch are the same. In a *double thread* screw, the lead is twice the pitch (Fig 12.2). The lead is equal to the pitch multiplied by an integral number.

- *The major diameter* is the largest diameter of the screw thread and *the minor diameter* is the smallest diameter of the screw thread Fig (12.3).

- *The depth of engagement* is the distance  measured radially over which a mating male and female thread overlap (Fig 12.4).

When a screw is used *without a nut* (as in a wood screw) it is usually called a *screw*, and when it is used *with a nut* it is called *a bolt*. Hence *bolts* and *nuts* belong together as a pair. Many types of screws are available with different *types of screw thread*, different *shapes of head,* different *types of nut*, etc. (Figs 12.5 & 12.6). *Locking devices* are quite often used to prevent nuts from becoming loose and some of these are shown in Fig 12.7.

### Keys

A key is a piece of steel inserted between a *shaft* and a *hub*, usually in an *axial direction*, to prevent rotation between them. *Recesses* called *keyways* are cut in both the shaft and the hub to accomodate the key. The commonest type of key is the *rectangular key* (Fig 12.8 (a)). It should fit tightly at the sides and have a *clearance* at the top. When the shaft and the hub have to move in *an axial direction* relative to each other, *a feather key* is used. This is a rectangular key which is fixed (usually by screws) to the hub or the shaft. A *working clearance* at the top and sides is necessary.

## Spline shafts and hubs

Spline shafts and hubs are used for heavy duty couplings. Such couplings allow an *axial movement* between the shaft and the hub. The spline shaft has a number of key like *projections* equally spaced round its circumference. These engage with corresponding *recesses* in the spline hub (Fig 12.8 (b)).

## Cotters

A cotter is a *flat wedge shaped* piece of steel which is used to fasten rods which are subjected to *axial forces only*. Slots are cut on the end of one rod and on a socket which is forged on the other rod. The wedge shaped cotter is forced into the two slots (Fig 12.8 (c)).

---

# Vocabulary

| | | | |
|---|---|---|---|
| advance | vorwärtsbewegen *v* | pitch | Teilung *f* |
| bolt and nut | Schraube mit Mutter *f* | projection | vorspringender Teil *m* |
| cone | Kegel *m* | recess | Einschnitt *m* |
| cotter | Keil *m* | ridge | Kamm *m* |
| cotter pin, split pin | Splint *m* | screw | Schraube *f* |
| heavy duty | hochleistungs.. *adj* | screw thread | Gewinde *n* |
| hub | Nabe *f* | shaft | Welle *f* |
| key | Passfeder *f* | slot | Schlitz *m* |
| keyway | Nut *f* | spline shaft | Keilwelle *f* |
| lead | Steigung *f* | | |
| locking device | Losdrehsicherung *f* | wedge | Keil *m* |

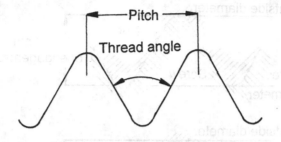

Fig 12.1 Angle of a screw thread

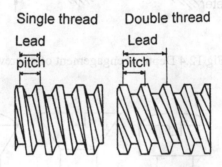

Fig 12.2 Relationship between lead and pitch for different screw threads

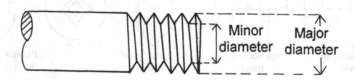

Fig 12.3 Major and minor diameters of a screw thread

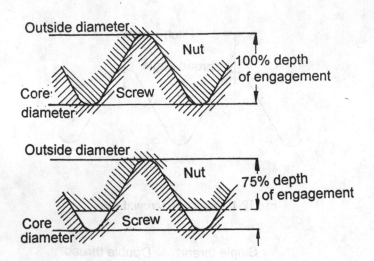

Fig 12.4 Depth of engagement of a screw thread

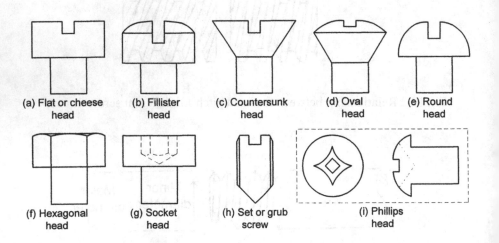

Fig 12.5 Some types of screw heads

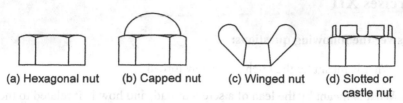

(a) Hexagonal nut    (b) Capped nut    (c) Winged nut    (d) Slotted or castle nut

Fig 12.6 Some types of nuts

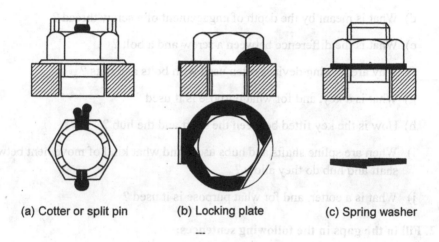

(a) Cotter or split pin    (b) Locking plate    (c) Spring washer

Fig 12.7 Some locking devices

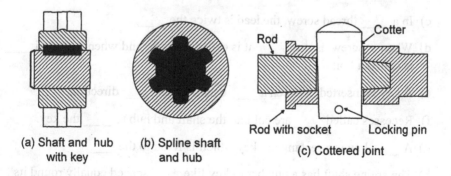

(a) Shaft and hub with key    (b) Spline shaft and hub    (c) Cottered joint

Fig 12.8 Keys, splines, and cotters

# Exercises XII

## 1. Answer the following questions:

a)  What is meant by the pitch of a screw thread ?

b)  What is meant by the lead of a screw thread, and how is it related to the pitch ?

c)  What are the major and minor diameters of a screw thread ?

d)  What is meant by the depth of engagement of a screw thread ?

e)  What is the difference between a screw and a bolt ?

f)  Why are locking devices often used with bolts and nuts ?

g)  What is a key, and for what purpose is it used ?

h)  How is the key fitted between the shaft and the hub ?

i)  When are spline shafts and hubs used, and what kind of movement between shaft and hub do they allow ?

j)  What is a cotter, and for what purpose is it used ?

## 2. Fill in the gaps in the following sentences:

a)  Screws are used for ____ various types of ____.

b)  The ____ is the distance advanced by the screw when rotated through ____.

c)  In a ____ thread screw, the lead is twice the ____.

d)  When a screw is used ____ it is called a screw, and when it is used ____ it is called a bolt.

e)  A key is inserted between a ____ usually in an ____ direction.

f)  Recesses called ____ are cut into the shaft and hub to ____ the key.

g)  A ____ key is a rectangular key which is fixed to the ____.

h)  The spline shaft has a number of key like ____ spaced equally round its ____.

    i) A cotter is used to _____ rods which are ___ axial forces only.

    j) It is necessary to use _____ devices to _____ nuts from becoming loose.

## 3. Translate into English:

    a) Ein Außengewinde wird auf der Außenseite eines Zylinders geschnitten, während ein Innengewinde auf der Innenseite einer Bohrung geschnitten wird, wie etwa bei einer Mutter.

    b) Es gibt viele Gewindearten, die jeweils für spezielle Einsätze ausgelegt sind. Jedes Gewinde wird nach genauen Vorgaben gefertigt und erhält einen eindeutigen Namen, wie z.B. Metrisches Gewinde.

    c) Bei einem eingängigen Gewinde sind die Teilung und Steigung gleich. Bei einem zweigängigen Gewinde ist die Steigung zweimal so groß wie die Teilung.

    d) Schrauben unterscheiden sich durch Gewindemaße, Kopfform und andere Einzelheiten. Sechskantschrauben sind die in Maschinenbau am häufigsten verwendeten Schrauben. Gewindestifte sind Schrauben ohne Kopf mit Gewinde auf der ganzen Schaftlänge.

## 13   Engineering inspection

Inspection of finished components is an essential part of all engineering production. If we consider the vast number of components produced by industry today, it is easy to see that it is not possible to inspect all of them for *dimensional accuracy*. With modern production methods it is possible to maintain a high degree of accuracy, but there are always *small variations* in dimensions which must be kept within *strict limits* to ensure the *interchangeability* of *components*.

### Limits and tolerances

When a component is designed, it is given certain dimensions on the technical drawing of the component which are called *nominal dimensions*. In the manufacturing process however, it is impossible to produce components that have precisely the dimensions stated in the drawing. For this reason, a small deviation from the nominal dimensions must be allowed, and this deviation is called a *tolerance*. When the component is manufactured, the dimension considered must lie between two values called the *high limit* and the *low limit*.

### Clearance fits

The essential conditions for a *clearance fit* (between a shaft and a hole) are shown in Fig 13.1. In this kind of fit, a shaft must be able to *move freely* in a hole without being too loose in it. It will be seen from the figure that both the shaft and the hole have *high* and *low limits*. If a clearance fit is to be ensured, then it is necessary that the high limit for the shaft, be smaller than the low limit for the hole. The tolerance for the shaft will be seen to be *above* and *below* the *nominal diameter*. Such a tolerance is called a *bilateral tolerance*. The tolerance for the hole however, will be seen to be *above* the *nominal diameter*. Such a tolerance is called a *unilateral tolerance*, and can be either above or below the nominal diameter.

### The engineering allowance

It has been mentioned before, that there is a small gap between the high limit of the shaft and the low limit of the hole. This quantity is known as an *allowance*. The size of the allowance determines the *quality of the fit*, where a high quality fit corresponds to a small allowance, and a low quality fit to a large allowance.

In general, allowances and tolerances must be kept as *large* as *possible*. Small tolerances increase the *cost* of *manufacture*, and also the number of *rejected components.*

## Interference fits

Sometimes the shaft is made very slightly larger than the hole. In this case, *considerable pressure* is *required* to force the shaft into the hole, and such a fit is called a *force fit* or an *interference fit*. This can be used as a cheap way of joining two components together permanently without using fasteners.

## Transition or push fits

If a shaft can be pushed into a hole by the use of hand pressure only, then this type of fit is called a *push fit* or a *transition fit*. A push fit should not be used if the shaft is to rotate. Slip bushes and dowels are good examples of push fits.

# Vocabulary

| | | | |
|---|---|---|---|
| **allowance** | Spiel *n*, Toleranz *f* | **inspect** | prüfen *v* |
| **bilateral** | zweiseitig *adj* | **interchangeable** | auswechselbar *adj* |
| **bush** | Lagerbuchse *f* | **measurement** | Messung *f* |
| **clearance** | Spielraum *m*, Spiel *n* | **mention** | erwähnen *v* |
| **consider** | überlegen, betrachten *v* | **nominal dimension** | Nennmaß *n* |
| **deviation** | Abweichung *f* | **quantity** | Größe *f* |
| **dimension** | Maß *n* | **reject** | ausscheiden *v* |
| **drawing** | technische Zeichnung *f* | **strict** | streng *adj* |
| **ensure** | sichern *v* | **tolerance** | Toleranz *f* |
| **fit** | Passung *f* | **transition (or push) fit** | Übergangspassung *f* |
| **freely** | zwanglos *adv* | **unilateral** | einseitig *adj* |
| **impossible** | unmöglich *adj* | **variation** | Veränderung *f* |
| **interfere** | überlagern *v* | | |

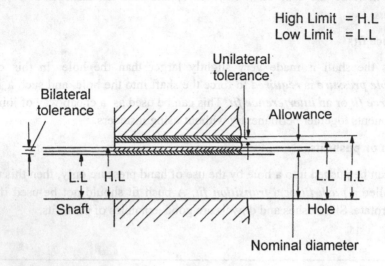

Fig 13.1 Essential conditions for a clearance fit

# Exercises XIII

## 1. Answer the following questions :

a) Why is engineering inspection so important ?

b) Why must variations in the dimensions of components be kept within strict limits?

c) What is meant by the nominal dimensions of a component ?

d) What is a tolerance, and why is it necessary to have this in the dimensions of a component?

e) What are the names of the two limits between which the dimensions of a component lie ?

f) What is meant by a clearance fit ?

g) Explain the meaning of the terms unilateral tolerance and bilateral tolerance.

h) What is meant by an engineering allowance ?

i) Why should tolerances and allowances be kept as large as possible ?

j) What is meant by an interference fit ?

## 2. Fill in the gaps in the following sentences:

a) Engineering _____ is an essential part of all manufacturing. However, it is not usually possible to inspect every component for _____ accuracy.

b) Small _____ in dimensions must be kept within strict _____.

c) Limits are _____ to ensure that components are ___.

d) The dimensions of a component given in the _____ are called _____.

e) It is _____ to produce components which have _____ the dimensions given in the _____.

f) A certain amount of _____ from the nominal dimensions must be _____ and this deviation is called a _____.

g) In a ____ fit, the shaft must be able to move freely in a ____ without being too ____ in it.

h) The engineering ____ is the difference between the ____ limit of the shaft and the low limit of the ____.

i) Small tolerances ____ the cost of manufacture, and also the number of ____ components.

j) If the shaft is made slightly ____ than the hole, pressure is required to ____ the shaft into the hole.

## 3. Translate into English:

a) Obwohl die technische Prüfung ein wesentlicher Teil der ganzen technischen Produktion ist, ist es nicht möglich, jedes hergestellte Teil zu prüfen.

b) Wenn ein Bauteil entworfen wird, so werden ihm bestimmte Abmessungen auf der technischen Zeichnung zugewiesen, welche Nennmaße genannt werden. Jedoch ist es unmöglich, eine große Anzahl von Bauteilen herzustellen, die präzise die Maße aufweisen, die in der Zeichnung angegeben sind.

c) Aus diesem Grund muss eine gewisse Abweichung von den Nennmaßen erlaubt sein, welche Toleranz genannt wird. Wenn das Bauteil hergestellt wird, muss jedes Maß zwischen zwei Werten liegen, die mit oberem und unterem Grenzmaß bezeichnet werden.

d) Um eine Welle in eine Bohrung zu fügen, muss es eine schmale Lücke zwischen dem oberen Grenzmaß der Welle und dem unteren Grenzmaß der Bohrung geben. Diese Lücke wird Spiel genannt.

e) Im allgemeinen müssen Toleranzen so groß wie möglich gehalten werden. Kleine Toleranzen erhöhen die Herstellungskosten und außerdem den Ausschuss.

# 14 Sheet metal cutting and blanking

Metal working presses are used by industry for the production of a large variety of articles from sheet metal. The *low cost* of producing *high quality metal pressings*, makes *press work* one of the most attractive and important of manufacturing processes. The first step in the manufacture of sheet metal products is called *shearing*. This is the process of *cutting a piece of metal* into the *required shape*. Some of the most commonly used metal shearing operations are given below.

## Types of shearing operations

1.  *Blanking* is the operation of cutting out a piece of metal of the desired shape by using a punch and a die. In this case, it is the *removed piece* of *metal* that is important and not the hole produced.

2.  *Piercing* is the operation of producing *a hole* of *any shape* in a sheet of metal using a punch and a die. The *material removed* is *unimportant*, and is treated as scrap. *Punching* is similar to piercing, but refers to the production of a *circular hole*.

3.  *Notching* is the operation of removing a piece of metal *from the edge* of a sheet of metal.

4.  *Lancing* is the operation of cutting through a metal sheet partially *without removing* any material.

5.  *Slitting* is the operation of cutting a sheet of metal in a straight line along its length.

6.  *Perforating* is the operation of producing a *regularly spaced pattern* of holes in a sheet of metal.

7.  *Nibbling* is the operation of cutting out a metal piece from a sheet of metal, by *repeated small cuts*.

8.  *Trimming* refers to the *removal* of *excess material* from a pressed object.

## Shearing processes using automatic presses

Large numbers of objects having the required shape can be produced rapidly using automatic presses. Two examples of the many possible ways in which this can be done are described below.

## Blanking

Blanking can be done by using a press that is fitted with an appropriate die set. The *double blanking die set* shown in Fig.14.1 consists of *two punches and a die*. The punches which have sharp edges at the bottom are attached to a punch holder which is fitted with *two guide collars.* The collars can move vertically on *two pillars* attached to the base of the die set. The punches are slightly smaller than the holes in the die.

The metal in the *form of strip* is first fed in from the right side up to the small *stop stud*. The punches now move down through the *stripper plate,* and *two blanks* are punched from the strip. As the punches rise after the downward stroke, the metal strip will be lifted, but is *prevented* from *rising* too far by the *stripper plate*. The metal strip is next pushed forward by an *appropriate distance* determined by the *position* of the *stop stud,* and the punches move down to produce two more blanks. This process is continued until the required number of blanks is produced.

## Combined blanking and piercing using a follow-on (or progressive) die set

A follow-on die set can be used to perform *two* or *more operations simultaneously* with *a single stroke* of the *press*. This is done by mounting separate sets of punches and dies in *two* or *more positions*. The metal sheet or strip is moved from one position to the other, until the complete part is produced.

A two position follow-on die set used for the *production of washers* is shown in Fig.14.2. The metal strip is fed in from the right into the first position where a hole is produced by the first die set in the first downward stroke of the ram. The metal strip is advanced to the next position, the *correct position* being controlled by the *stop* shown in the diagram.In the second stroke of the ram, the *pilot* enters the already *pierced hole* and *locates* it correctly, while the blanking punch moves down and shears the metal strip to produce a washer.

Although two strokes of the ram are needed to produce a washer, a *complete washer* is delivered after *each stroke* of the ram due to the follow-on action.

## The shearing of metal by hand and by using simple machines

Thin metal sheets can be cut by hand using *hand shears (or snips),* while thicker sheets can be cut using *bench shears. Guillotines* can be used to cut larger sheets, and *band saws* can be used to cut *complex shapes.*

# Vocabulary

| | | | |
|---|---|---|---|
| **advance** | vorankommen *v* | **pierce** | lochen *v* |
| **appropriate** | geeignet *adj* | **pillar** | Säule *f* |
| **ascend** | ansteigen *v* | **pilot** | Suchstift *m* |
| **attractive** | anziehend *adj* | **press** | Presse *f* |
| **bead** | bördeln, falzen *v* | **pressing** | gepresstes Metallstück *n* |
| **blank** | ausgeschnittenes Metallstück *n* | **progressive** | schrittweise *adj* |
| **blank** | ausschneiden *v* | **punch** | Schneidstempel *m* |
| **collar** | Muffe *f* | **punch** | stanzen, lochen *v* |
| **cutting edge** | Schneidkante *f* | **ram** | Stößel *m* |
| **die block** | Schneidplatte *f* | **require** | erfordern, brauchen *v* |
| **die set** | Schneidwerkzeug *n* | **shear** | scheren *v* |
| **downwards** | abwärts *adv* | **shears** | Scherer *m* |
| **ensure** | sicherstellen *v* | **slit** | schlitzen *v* |
| **follow-on die set** | Folgeschneidwerkzeug *n* | **strip** | Streifen *m* |
| **holder** | Behälter *m* | **stripper plate** | Abstreifer *m* |
| **involve** | verwickeln *v* | **stroke** | Hub *m* |
| **lance** | einschneiden *v* | **stud** | Stift *m* |
| **nibble** | nibbeln, knabberschneiden *v* | **trim** | trimmen *v* |
| **notch** | ausklinken *v* | **washer** | Unterlegscheibe *f* |
| **perforate** | lochen, perforieren *v* | | |

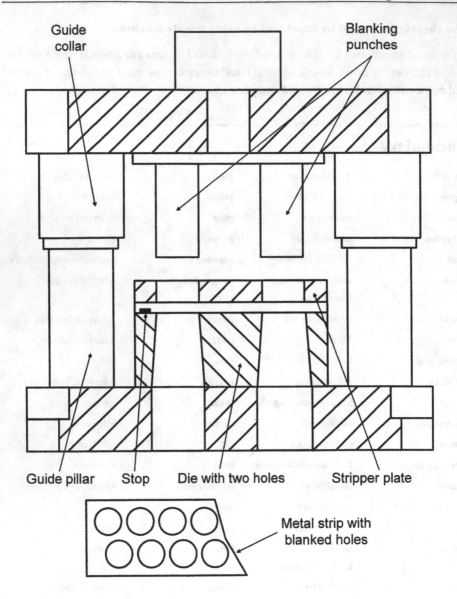

Fig 14.1 Blanking die set with two punches

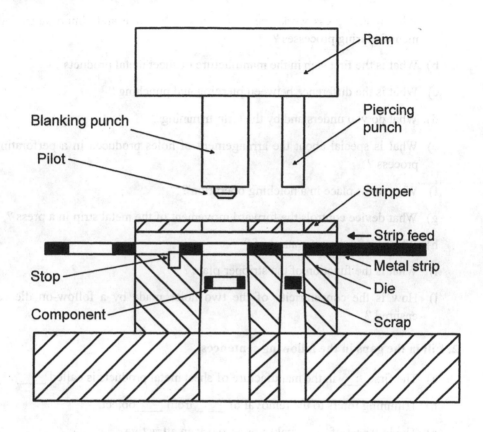

Ram

Blanking punch

Piercing punch

Pilot

Stripper

Strip feed

Metal strip

Stop

Die

Component

Scrap

Fig 14.2 Combined blanking and piercing using a follow-on die set

# Exercises XIV

## 1. Answer the following questions:

a) What makes press work one of the most attractive and important of all manufacturing processes ?

b) What is the first step in the manufacture of sheet metal products ?

c) What is the difference between piercing and punching ?

d) What do you understand by the term trimming ?

e) What is special about the arrangement of holes produced in a perforating process ?

f) What takes place in a notching operation ?

g) What device controls the forward movement of the metal strip in a press ?

h) What does a die set consist of ?

i) What is the function of the stripper plate ?

j) How is the concentricity of the two holes made by a follow-on die set ensured ?

## 2. Fill in the gaps in the following sentences:

a) The first _____ in the manufacture of sheet metal products is called _____ .

b) Trimming refers to the removal of _____ from _____ object.

c) The low cost of _____ makes press work an attractive _____ .

d) Piercing is the _____ of producing a hole of _____ in a sheet of metal.

e) Notching is the operation of _____ a piece of metal from the _____ .

f) Perforating is the operation of producing a _____ pattern of _____ in a sheet of metal.

g) The blanking _____ carries a pilot which fits into the _____ hole.

h) Blanking can be done _____ a press fitted with _____ .

i)  The punches are slightly ____ than the holes ____.

j)  The metal strip is ____ from rising by the ____.

## 3. Translate into English:

a)  Metallpressen werden in der Industrie für die Produktion einer Vielzahl von Gegenständen aus Metallblech verwendet. Die niedrigen Kosten und die hohe Qualität der Metallpressstücke, die von der Industrie hergestellt werden können, machen diesen Prozess sehr attraktiv.

b)  Das Ausschneiden von Gegenständen aus Metallblech kann durch den Einsatz einer mit geigneten Schneidwerkzeugen ausgerüsteten Metallpresse bewerkstelligt werden. Das in der Zeichnung dargestellte Schneidwerkzeug besteht aus zwei Schneidstempeln und einem Schneideisen. Der Metallstreifen wird von rechts bis zur kleinen Stoppschraube zugeführt. Die Schneidstempel bewegen sich abwärts durch den Abstreifer und stanzen zwei Stücke aus den Streifen.

c)  Lochen ist der Vorgang des Herstellens eines Lochs beliebiger Größe in einem Metallblech unter Benutzung eines Schneidstempels und eines Schneideisens. Das entfernte Material ist unwichtig und wird als Abfall behandelt, im Gegensatz zum Ausschneiden, wo das entfernte Materialstück der erwünschte Gegenstand ist.

d)  Das Ausschneiden ist der Vorgang des Ausschneidens eines Stücks Metall der gewünschten Form unter Verwendung eines Schneidstempels und eines Schneideisens. Die Perforation ist der Vorgang der Herstellung eines Musters von Löchern in gleichmäßigem Abstand auf einem Metallblech.

# 15 Sheet metal bending and forming processes

The bending and forming of metals are *cold working* operations which involve *plastic deformation* of the workpiece. The metal has to be stretched beyond the *elastic limit*, but not so far that it *cracks* or *fractures*. Only some metals and alloys have sufficient *ductility* to be used in this way. Low carbon steels, killed steels, and alloys of copper and aluminium are widely used for the manufacture of *metal pressings*.

## Bending

Many metal components are formed by bending a sheet of metal in one or more places. Short lengths of metal can be bent by using die sets in mechanical presses. Longer lengths require special presses with *long beds* called *press brakes*. The kind of *die set* used in a press brake is shown in Fig 15.1. Complex shapes can be formed by *repeated bending* as shown in Fig 15.2. Complex profiles can also be formed by passing metal sheets or strips through *successive sets* of *rollers*.

### Stretch forming and stretch drawing

In pure *stretch forming*, the metal sheet is completely *clamped* round its *circumference* and the change in shape is achieved at the expense of sheet *thickness*. Stretching the metal causes it to become thinner. The advantage of such a process is that only a *single punch* needs to be used to stretch a sheet which is clamped using a number of clamps as shown in Fig 15.3.

An example of *stretch drawing* is shown in Fig 15.4. The metal blank is clamped round its circumference by *a blank holder*. The punch moves downwards, stretching and drawing the metal into the cavity in the die.

### Deep drawing

In the deep drawing process, the metal blank is *not clamped* but allowed *to draw* into *the die*. Deep drawing is usually done in a number of stages. The first stage which is also called *cupping* is shown in Fig 15.5.

Initially the blank is held firmly on the die by the *pressure pad*. The punch moves downwards, and pushes the blank into the cavity. The metal is made to bend and flow plastically while it is drawn over the edges of hole to form a cup. The thickness is very little changed. The pressure pad has the function of *ironing out* any *wrinkles* formed during the drawing process. It does not however prevent the metal from being drawn into the cavity. Deeper objects can be produced by *redrawing several times.*

*Combination dies* can be used to carry out blanking and drawing simultaneously.

## Some other press forming operations

- *Beading* is an operation in which the edge of a metal sheet is *folded over* to improve its strength, stiffness, safety, and appearance (Fig 15.6).

- *Plunging* is an operation in which a punch is pressed through a hole in a metal sheet, bending it into the shape required to take the head of a screw (Fig 15.7).

- *Flanging* is an operation which produces *edges* of various *widths* and *angles* on flat or curved metal sheets and tubes (Fig 15.8).

- *Coining* is a process in which a a very high pressure is applied from both sides on a piece of metal placed between a punch and a die. The *metal flows* in the *cold state* and fills up the cavity between the punch and the die.

- *Embossing* is similar to coining, and involves *plastic flow* of the metal. It is an operation in which figures, letters, or designs are formed on sheet metal parts.

## Spinning

In the spinning process, a thin sheet of metal is formed into the required shape by *revolving* it at *high speed,* and *pressing it* against a *former* attached to the headstock spindle of a lathe. The metal is also supported at the tailstock. Pressure is applied by a *special tool*, forcing it to acquire the *shape* of the *former* (Fig 15.9).

# Vocabulary

| | | | |
|---|---|---|---|
| **bead** | falzen, bördeln *v* | **impression** | Prägung *f* |
| **bend** | biegen *v* | **prevent** | verhindern *v* |
| **cavity** | Hohlraum *m* | **press** | Presse *f* |
| **circumference** | Umkreis *m* | **press** | drücken, pressen *v* |
| **crack** | Riss *m*, Spalt *m* | **pressure pad** | Niederhalter *m* |
| **deep draw** | tiefziehen *v* | **revolve** | sich drehen *v* |
| **elastic limit** | elastische Grenze *f* | **spin** | drücken, spinnen *v* |
| **emboss** | prägen *v* | **squeeze** | auspressen *v* |
| **engrave** | gravieren *v* | **stiffness** | Steifheit *f* |
| **flange** | bördeln *v* | **stretch form** | streckziehen *v* |
| **form** | formen, bilden *v* | **tailstock** | Reitstock *m* |
| **headstock** | Spindlestock *m* | **wrinkle** | Falte *f* |

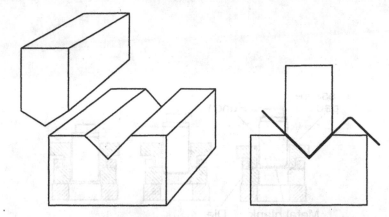

Fig 15.1 Die set for a press brake     Fig 15.2 Repeated bending
                                                      using a press brake

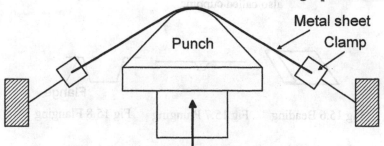

Fig 15.3 Stretching forming using a punch without a die

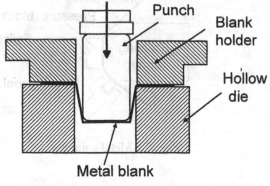

Fig 15.4 Stretch drawing

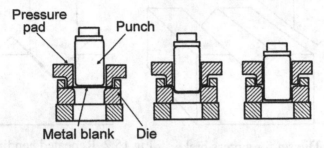

Pressure pad    Punch

Metal blank    Die

Fig 15.5 First stage in a deep drawing process,
also called cupping

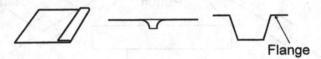

Fig 15.6 Beading    Fig 15.7 Plunging    Fig 15.8 Flanging

Flange

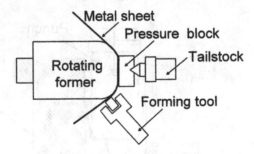

Metal sheet
Pressure block
Tailstock
Rotating former
Forming tool

Fig 15.9 Metal spinning

# Exercises XV

## 1.Answer the following questions:

a)  What kind of deformation does a metal undergo in a bending process ?

b)  What kind of metals can be bent easily without breaking ?

c)  What kind of presses are required to bend long lengths of metal ?

d)  How is the metal sheet clamped in the stretch forming process ?

e)  What kind of change in thickness does a metal sheet experience in the deep drawing process ?

f)  What is the function of a pressure pad ?

g)  What are combination dies used for ?

h)  What is a plunging operation ?

i)  What kind of operation is embossing ?

j)  How far is a metal stretched in bending and  forming processes ?

## 2. Fill in the gaps in the following sentences:

a)  The bending and _____ of metals are _____ operations.

b)  The metal has to be _____ beyond the _____.

c)  Only some metals have sufficient _____ to be used in _____.

d)  Complex _____ can also be formed by passing the metal strip through _____.

e)  In pure _____ forming the metal sheet is completely _____ round its circumference.

f)  The _____ is held firmly on the die by the _____.

g)  The punch _____ and pushes the blank into the _____.

h)  In the _____, a thin sheet of metal is formed into the _____ by _____at high speed and pressing it against a former.

i)  Short lengths of metals can be _____ using die sets in _____.

    j) Longer lengths require special presses with ____ called ____.

**3. Translate into English:**

    a) Das Metall muss bis jenseits der elastischen Grenze gedehnt werden. Lediglich einige Metalle und Legierungen haben eine ausreichende Dehnbarkeit, um in dieser Weise genutzt zu werden.

    b) Kurze Metallstücke können unter Verwendung von Stempeln in mechanischen Pressen gebogen werden. Längere Metallstücke erfordern Spezialpressen mit besonders langen Bänken. Komplexe Formen können durch wiederholtes Biegen geformt werden.

    c) Beim Streckziehen wird das Metall vollständig um seinen gesamten Umfang eingespannt. Durch das Ziehen wird das Metall dünner. Die Veränderung der Form wird auf Kosten der Blechstärke erzielt.

    d) Im Prozess des Drückens rotiert ein dünnes Metallblech mit hoher Geschwindigkeit, während ein spezielles Werkzeug es gegen einen am Spindelstock der Drehbank befestigten Former drückt. Das Metallblech wird auch von dem Reitstock gestützt. Der vom Werkzeug ausgeübte Druck zwingt das Metall, die Form des Formers anzunehmen.

# 16  The lathe and single point cutting tools

*The lathe* is a machine tool which is used for producing components that are symmetrical about an axis. It can be used for machining *cylindrical surfaces*, both external and internal, and also for the turning of *conical surfaces* or *tapers*. An additional feature of the lathe, is its ability to cut *screw threads* on a cylindrical surface that has already been machined. The accuracy of the work done on a lathe depends on the *skill* and *experience* of the operator. A lot of time is taken for *tool changing, tool setting,* etc. with the result that it is unsuitable for production work. It is mainly used in tool rooms and workshops for the making of prototypes, and also for maintenance work.

Fig 16.1 shows the different parts of a typical lathe. It has *a rigid bed* with *parallel guideways* on which are mounted a *fixed headstock* and a *movable tailstock.* In addition there is a *carriage* which can be moved along the guideways of the bed, in a direction which is parallel to the axis of rotation of the *headstock spindle.* The headstock has a strong spindle which is driven by an electric motor through a *gearbox.* The speed of the spindle can be varied through a wide range to suit the type of work that has to be done. A lathe is also usually fitted with a *leadscrew,* which can be geared to the headstock spindle through the gearbox.

The work to be machined is held in a device called a *chuck,* which is fitted to the *nose* or *front end* of the spindle. As the spindle rotates the work rotates with it, and can be turned down to the desired dimensions using *single point tools* which are held in a *tool post.* The toolpost is *fixed* on the *carriage,* and the tool can be moved *parallel* to the axis, or at an *angle* to it. *Screw threads* can be cut by coupling the carriage to the *leadscrew.* The lathe is a versatile machine with a wide range of accessories, and can be used to produce a large variety of components.

## Capstan and turret lathes

Although the ordinary lathe is unsuitable for *mass production* work, modified forms of the lathe like *capstan* and *turret* lathes have been used as mass production machines. These lathes have the same type of headstock and fourway tool post as the ordinary lathe. The tailstock is however replaced by a *hexagonal turret*, and each face of the turret   can carry one or more tools. These tools may be used successively to perform different operations in a regular sequence.

The *feed movement* of each tool can be *regulated* by *stops*. Lead screws are not usually fitted to these lathes, and screw cutting is usually done by using taps and dies. Capstan and turret lathes have different types of turrets as shown in Fig. 16.2 and Fig 16.3. The *tool setting* has to be done by a *skilled operator*, but once this has been done, a *semi-skilled operator* can produce  a large number of components in a short time. The tool turret is also used in *CNC machines*  (see Chapter 20).

## Single point cutting tools

Many types of metal cutting machines like lathes and shaping machines use  single point metal cutting tools. For efficient metal cutting, it is absolutely important that the correct *cutting angles* be ground on the tip of the tool. The values of the angles depend on the material being cut. Angles that are important in a lathe cutting tool are the following: *Front* or *top rake angle A, Side rake angle B, Front clearance angle C, Side clearance angle D* (Fig 16.4).

## Carbide tipped tools

Fast cutting speeds involve considerable metal removal in a minimum of time.  It is essential that tool wear be kept to a minimum, if the dimensional accuracy of the components being produced is to be maintained. *Carbide tipped tools* are almost as hard as *diamond*, and will maintain a sharp cutting edge under conditions that would cause ordinary tools to burn away. Carbide is the name given to several alloys of carbon and some metals like *tungsten*, *titanium*, and *tantalum.* Carbides are brittle materials, and can only be used as tips for tools. A *carbide tip* is usually *brazed* on a tool body made of ordinary steel, and then *ground* to the *required shape.*

## Lubrication and the use of cutting fluids

When fast cutting speeds are used, a small amount of cutting fluid is pumped over the cutting edge of the tool. The *cutting fluid* removes the heat generated during the cutting process, and thus increases the life of the tool. It also helps to remove chips from the edge of the tool, and improve the surface finish. *Water soluble oils* are cheap and efficient coolants and are suitable for the machining of most steel components. However, they *do not* have very good *lubricating properties* and are unsuitable for more complex machining operations such as gear cutting.

# Vocabulary

| | | | |
|---|---|---|---|
| ability | Fähigkeit *f* | metal chips | Metallspäne *m* |
| absolutely | vollkommen *adv* | perform | leisten *v* |
| accessory | Zubehörteil *n* | practice | Gewohnheit f |
| application | Anwendung *f* | precision | Genauigkeit *f* |
| bed | Bett *n* | rake angle | Spanwinkel *m* |
| capstan lathe | Revolverdrehbank *f* | range | Reihe *f*, Bereich *m* |
| carriage | Werkzeugschlitten *m* | reduce | verkleinern *v* |
| change | wechseln *v* | regulate | regeln *v* |
| chuck | Futter *n* | rigid | stabil *adj* |
| clearance angle | Freiwinkel *m* | semi-skilled | angelernt *adj* |
| conical | kegelförmig *adj* | sequence | Reihenfolge *f* |
| coolant | Kühlschmiermittel *n* | sideways | seitwarts *adv* |
| experience | Erfahrung *f* | similar | ähnlich *adj* |
| fluid | Flüssigkeit *f* | single point tool | Schneidmeißel *m* |
| four way tool post | Vierfachmeißelhalter m | skill | Geschicklichkeit *f* |
| groove | Nut *f* | successively | hintereinander *adv* |
| guideways | Bettführungen *pl* | surface finish | Oberflächenqualität f |
| headstock | Spindelstock *m* | tailstock | Reitstock *m* |
| involve | verwickeln *v* | tip | Spitze *f* |
| lathe | Drehmaschine *f* | tool post | Meißelhalter *m* |
| leadscrew | Leitspindel *f* | turret | Drehkopf *m* |
| machine tool | Werkzeugmaschine *f* | versatile | vielseitig *adj* |
| mass produce | serienmäßig herstellen *v* | | |

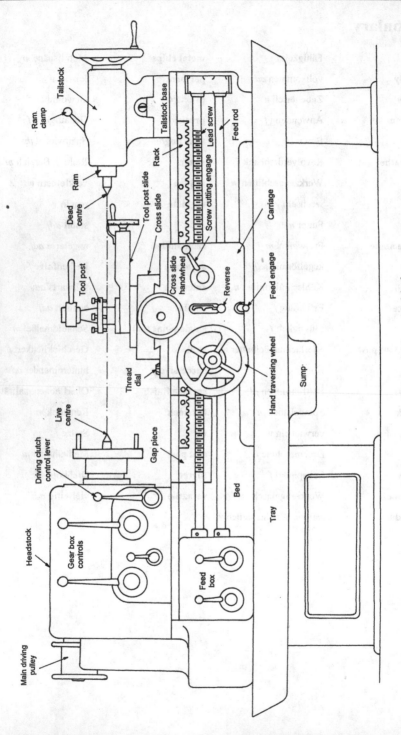

Fig 16.1 Parts of a typical lathe

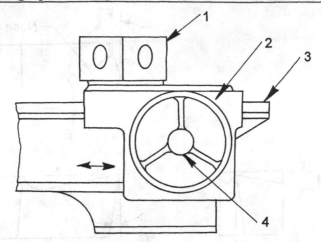

1. Hexagonal turret, 2. Turret saddle, 3. Lathe bed, 4. Handwheel for saddle

Fig 16.2 Turret lathe saddle and components

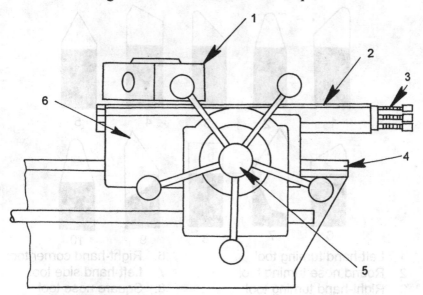

1. Hexagonal turret, 2. Auxilliary slide, 3. Feed stop rod, 4. Lathe bed, 5.Handwheel for auxilliary slide, 6.Saddle

Fig 16.3 Capstan lathe saddle and components

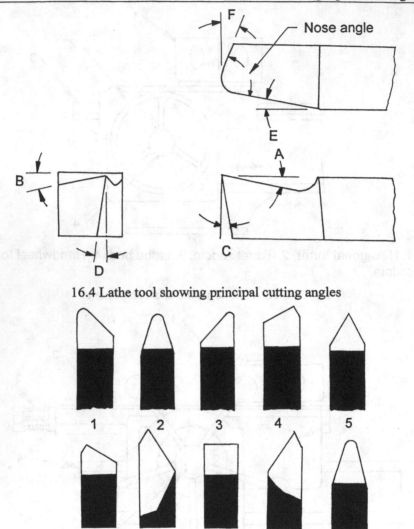

16.4 Lathe tool showing principal cutting angles

1.  Left-hand turning tool
2.  Round nose turning tool
3.  Right-hand turning tool
4.  Left-hand corner tool
5.  Threading tool

6.  Right-hand corner tool
7.  Left-hand side tool
8.  Square nose tool
9.  Right-hand side tool
10. Brass tool

Fig 16.5 A set of lathe tools suitable for different operations

# Exercises XVI

## 1. Answer the following questions:

a) What does the accuracy of the work done on a lathe depend on ?

b) Why is a lathe unsuitable for production work ?

c) What are the main parts of a lathe ?

d) What is the main component of a headstock, and how is it driven ?

e) How is the workpiece usually held on a lathe  ?

f) What is the purpose of the toolpost, and where  is it fixed ?

g) In what ways are capstan and turret lathes similar to the ordinary lathe, and in what ways are they different ?

h) What are the important angles in a lathe tool ?

i) Why is a cutting fluid used ?

j) What are the advantages and disadvantages of water soluble oils as coolants ?

## 2. Fill in the gaps in the following sentences:

a) Lathes can _____ components that are _____ about an axis.

b) The accuracy of the work done on a lathe depends on the _____ and _____ of the operator.

c) The ordinary lathe is unsuitable for _____ because a lot of time is spent on _____.

d) The lathe has a bed on which are mounted a fixed _____ and a movable _____.

e) The work to be _____ is held in a _____.

f) The work is _____ down to the desired _____.

g) The headstock contains a strong _____ driven by an electric motor through a _____.

h) Many machine tools like lathes and _____ use _____ tools.

i) The cutting fluid helps to remove _____ from the edge of the tool, and improves the _____.

j) It is absolutely important that the _____ angles be ground on the tip_____ .

## 3. Translate into English .

a) Eine Drehmaschine kann dazu benutzt werden, zylindrische Oberflächen, sowohl innen als auch außen und auch konische Oberflächen zu drehen. Ein weiteres Merkmal der Drehmaschine ist die Fähigkeit, Gewinde auf zylindrische Oberflächen zu schneiden.

b) Die Genauigkeit der Arbeit, die auf der Drehmaschine durchgeführt wird, hängt von der Geschicklichkeit und der Erfahrung des Drehers ab. Es wird viel Zeit für das Einrichten des Werkzeuges, das Werkzeugwechseln etc. benötigt, mit dem Ergebnis, dass diese Arbeit nicht für die Produktion geeignet ist.

c) Die Drehmaschine hat ein stabiles Bett mit parallelen Führungen, auf dem ein fester Spindelstock und ein beweglicher Reitstock angebracht sind. Außerdem gibt es einen Werkzeugschlitten, welcher entlang der Führungen in einer Richtung parallel zu der Achse der Drehbewegung der Spindel bewegt werden kann.

d) Der Gebrauch eines Kühlschmiermittels entfernt die Hitze, die während des Schneidvorganges erzeugt wird, und erhöht somit die Lebensdauer des Werkzeuges. Das Kühlschmiermittel dient auch dazu, Späne von der Schneidkante des Werkzeuges zu entfernen.

# 17 Drilling, milling, and broaching

A hole in a large mechanical component is usually produced by using a drilling machine. *Drilling machines* can be used for many types of operations such as,

*(1) Drilling, (2) Reaming, (3) Boring, (4) Counterboring, (5) Countersinking, (6) Tapping, (7) Grinding, (8) Trepanning.*

The drilling of a hole of the right size in the right position can be a difficult operation. In mass production, *jigs and fixtures* are often used to ensure the *accurate location* of holes in a component. Very precise drilling can be achieved by using a machine called a *jig borer*. Here the work is clamped on a *compound table* which has two movements in two directions at right angles to each other. The work is moved by using the *lead screws* on the compound table until the position of the hole is *precisely located* under the *drilling head*, after which the drilling can be done. It can also be done less precisely, by marking the location by hand with a *centre punch* and drilling at this location. A *centre drill* (Fig 17.5) is first used to drill a small hole, after which a *twist drill* is used to enlarge the hole to the required size.

**Twist drills**

A twist drill is made from a cylindrical piece of high speed steel. The three main parts of a twist drill are *(a) the body* which is the cutting unit, *(b) the shank* which is the part gripped by the drilling machine chuck, and *(c) the tang* which is found only in large *tapered shank* drills (Fig 17.1).

Small drills have *straight shanks* which are held in a *self-centering chuck*. Large drills usually have tapered shanks with a tang, and are inserted directly into the spindle of the machine. When a drill with a small tapered shank is used in a machine with a large socket, a *matching sleeve* is required (Fig 17.2). Holes produced by a twist drill are *slightly oversize*, and have a *rough inner surface*.

*Reaming* is an accurate way of finishing a hole to the right size. The hole is usually drilled slightly *undersize,* and then enlarged to the correct size using a *reamer* which removes only a *small amount* of metal. A reamer has *flutes* and *multiple cutting* edges which give the hole a *smooth finish* (Fig17.3).

*Counterboring* enlarges the end of a hole cylindrically to accomodate the heads of bolts. The tool used is called a *counterbore* (Fig 17.4).

*Countersinking* enlarges the end of a hole conically to accomodate the head of a countersunk screw or rivet. The tool used is called a *countersink* (Fig 17.6).

## Milling machines

The milling machine uses a *multipoint cutting tool* instead of a *single point cutting tool*. The use of multipoint cutting tools enables the milling machine to achieve fast rates of metal removal and also produce a good surface finish.

## Column and knee type of milling machines

The commonest types of milling machine have a main frame or *column*, and a projecting *knee* which carries the *saddle* and the *work table*. These machines are extremely versatile.They have three *independent* movements of the work table, *longitudinal, transverse*, and *vertical*. Milling machines are used in tool rooms and workshops, but they lack the *rigidity* required for heavy production work. There are three types of column and knee milling machines, (1) the horizontal milling machine, (2) the vertical milling machine, (3) the universal milling machine.

## The horizontal milling machine

A horizontal milling machine (Fig 17.7) has a *horizontal spindle* which is located in the upper part of the column. It receives power from the motor through *belts*, *gears,* and *clutches.* The spindle projects slightly out of the column face, and has a tapered hole into which *cutting tools* and *arbors* may be inserted. An arbor is an *extension* of the machine spindle on which milling cutters can be mounted. The *overhanging arm* which is fixed on the top of the column, serves as a bearing support for the arbor. The arbor has a taper shank which fits into the nose of the machine spindle. Fig 19.8 shows a spindle and arbor assembly with a mounted cutter.

## The vertical milling machine

The vertical milling machine has a column and knee similar to a horizontal milling machine but the spindle is *perpendicular* to the *work table*. The spindle head which is clamped to the vertical column may be *swivelled* at *an angle*, thus permitting the milling of angular surfaces.

## The universal milling machine

The universal milling machine is a versatile machine which can perform a wide variety of operations. It has a *fourth table movement* in addition to the three movements mentioned before. The table can be *swivelled* and moved at an angle to the milling machine spindle. It is also provided with accessories like *dividing heads, vertical milling attachments, rotary tables*, etc. These accessories enable the machine to produce *spur, spiral,* and *bevel gears, twist drills, milling cutters*, and to do a variety of *milling* and *drilling* operations.

## Broaching

Broaching is a method of metal cutting in which a suitably profiled multiple-edged cutting tool called a *broach* is *pushed* or *pulled* along the inner or outer surfaces of a work piece. Only a few mm of metal can be removed by this process, and it is therefore necessary that most of the metal removal has been done previously by other machining processes. Fig 17.9 shows the parts of an internal pull broach.

Broaching is carried out when unusual profiles with good surface quality and high dimensional accuracy are required. Only a single operation is required and high production rates using unskilled labour are possible.

*Internal broaching* is used to enlarge and cut holes of various contours in cylindrical holes made by drilling, casting, forging, etc. For example a broach may be used to cut splines or keyways in a previously drilled cylindrical hole. Broaching of *internal spirals* is also possible by using special broaches and rotating the workpiece, in a way similar to that in which screw threads are cut using taps.

*External broaching* can be used to produce external surfaces having a variety of profiles similar to those produced by milling, planing, shaping, etc.

# Vocabulary

| | | | |
|---|---|---|---|
| accessory | Zubehör n | movement | Bewegung f |
| arbor | Fräserdorn m | multipoint cutting tool | mehrschneidiges Werkzeug n |
| attachment | Zusatzgerät n | overhanging arm | Gegenhalter m |
| broaching | räumen v | permit | erlauben v |
| clutch | Kupplung f | project out | vorspringen v |
| compound table | Rechtecktisch m | reamer | Reibahle f |
| column | Säule f | rotary table | 360°drehbarer Schraubstock m |
| counterbore | zylindrische Senkung f | self-centering chuck | Dreibackenbohrfutter n |
| countersink | Kegelsenkung f | sleeve | Muffe f |
| dividing head | Universal Teilkopf m | spindle | Spindel f |
| drift | Austreiber m | support | Träger m |
| drill | Bohrer m | swivel | schwenken, drehen v |
| flute | Spannut f | tang | Kegellappen m |
| grind | schleifen v | taper shank | Kegelschaft m |
| knee | Konsole f, Knie n | transverse | quer adj |
| longitudinal | längs adj | versatile | vielseitig adj |

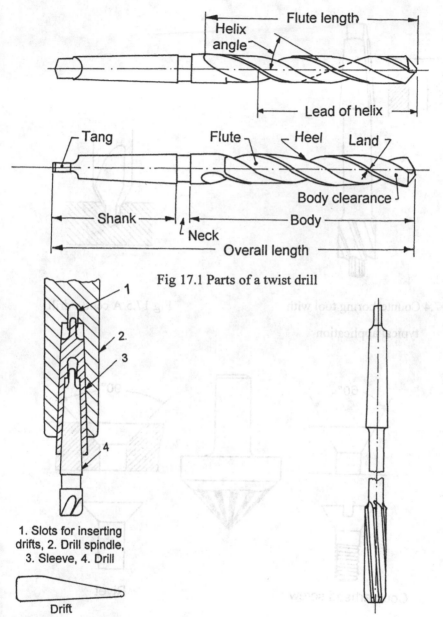

Fig 17.1 Parts of a twist drill

1. Slots for inserting
drifts, 2. Drill spindle,
3. Sleeve, 4. Drill

Drift

Fig 17.2 Drill, sleeve, and spindle                    Fig 17.3 One type of machine reamer

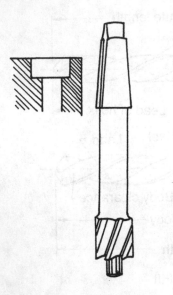

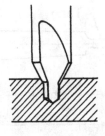

Fig 17.4 Counterboring tool with

typical application

Fig 17.5 A centre drill

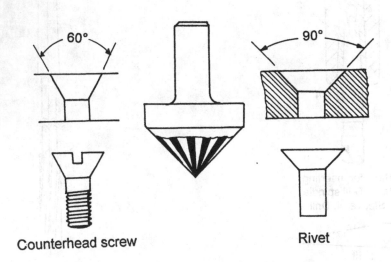

Counterhead screw                                    Rivet

Fig 17.6 Countersinking tool with typical application

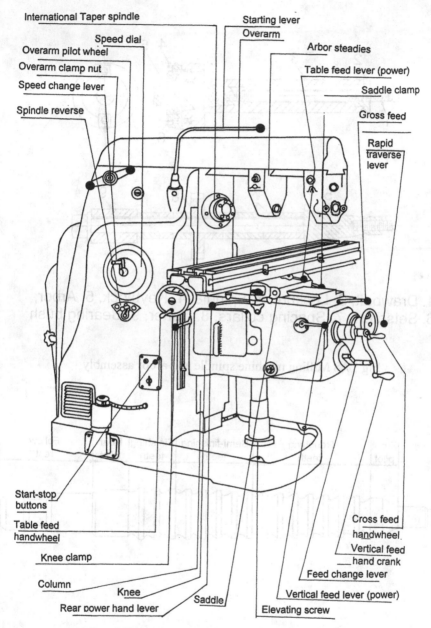

Fig 17.7 A horizontal milling machine

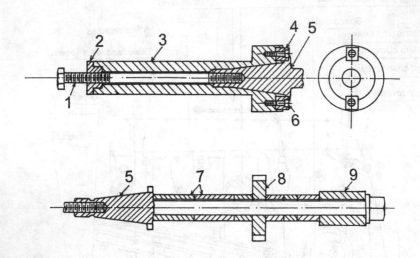

1. Drawbolt, 2. Locknut, 3. Spindle, 4. Keyblock, 5. Arbor,
6. Setscrew, 7. Spacing collars, 8. Cutter, 9. Bearing bush

Fig 17.8 Milling machine spindle and arbor assembly

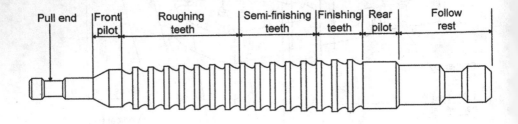

Fig 17.9 Broach suitable for internal pull broaching

# Exercises XVII

## 1. Answer the following questions:

a) What is a compound table and how is it used to drill a hole in a given location ?

b) What are the different parts of a twist drill ?

c) How are small drills held in a drilling machine ?

d) What is a reamer used for ?

e) What is meant by counterboring and countersinking ?

f) What is the difference between the cutting tools used in a lathe and those used in a milling machine ?

g) What kind of spindle head does a vertical milling machine have ?

h) Why are ordinary milling machines unsuitable for heavy work ?

i) What additional accessories can be used with a universal milling machine ?

j) What type of work can be done using these accessories ?

## 2. Fill in the gaps in the following sentences:

a) A compound table has two _____ movements at _____ angles to each _____.

b) Small drills have _____ shanks and are held in a _____ chuck.

c) Larger drills have _____ shanks and are fitted directly into the _____ of the drilling machine.

d) A hole can be enlarged _____ to the correct _____ by using a _____.

e) Counterboring _____ the end of a hole _____.

f) Milling machines use _____ cutting tools and not _____ cutting tools.

g) Cutting tools and _____ can be _____into a hole in the spindle.

h) The spindle _____ power from the motor through belts, _____ and clutches.

i)  The use of multipoint cutting tools ____ the milling machine  to ____ fast
    rates of ____.

j)  Milling machines are ____ with accessories like ____ heads and ____
    tables.

## 3. Translate into English

a)  Ein Bohrer wird gewöhnlich aus einem zylindrischen Stück eines HSS-
    Stahles gefertigt und hat spiralförmige Rillen. Am Anfang eines Bohr-
    vorganges wird ein Zentrierbohrer benutzt, um ein kleines Loch zu bohren.
    Dann wird ein Spiralbohrer eingesetzt, um das Loch in der richtigen Größe
    zu bohren.

b)  Kleine Bohrer haben gewöhnlich zylindrische Schäfte und werden in
    Dreibacken-Spannfuttern gehalten. Größere Bohrer haben gewöhnlich
    kegelförmige Schäfte, welche es ermöglichen, den Bohrer in die Spindel der
    Maschine einzusetzen.

c)  Diese Maschinen sind sehr vielseitig und haben drei unabhängige
    Tischbewegungen - längs, quer und vertikal. Sie werden in Werkstätten
    eingesetzt, aber es mangelt an Stabilität, die für schwere Arbeit in der
    Produktion erforderlich ist.

d)  Die Spindel hat eine kegelförmige Bohrung, in die verschiedene
    Schneidwerkzeuge und Fräsdorne eingesetzt werden können. Ein Fräsdorn
    erweitert die Fräsmaschinenspindel, und Schneidwerkzeuge können daran
    befestigt werden.

e)  Bei einer Senkrecht-Fräsmaschine ist der Spindelkopf senkrecht zu dem
    Arbeitstisch. Der Spindelkopf kann auch um einen Winkel gedreht werden,
    um somit das Fräsen von winkligen Oberflächen zu ermöglichen.

## 18 Surface finishing processes

### Grinding

*Grinding* is an operation performed by a rotating *abrasive wheel,* which removes metal from the surface of an object. It is usually a *finishing operation* which gives a good *surface finish* and high *dimensional accuracy* to workpieces which have already been machined by other methods. Very little metal is removed in this operation. Grinding is also used to machine materials which are too hard to be machined by other methods. The different types of grinding processes are:

*(1) External cylindrical grinding*, including *centreless grinding (2) Internal cylindrical grinding (3) Surface grinding (4) Form grinding.*

*External cylindrical grinding* is used to produce a *cylindrical* or *tapered surface* on the outside of a workpiece. The workpiece is rotated about its own axis as it moves lengthwise while in contact with a revolving grinding wheel.

*Internal cylindrical grinding* is used to produce *cylindrical holes* or *internal tapers* on a workpiece. The workpieces are rotated about their own axis, while the grinding wheel rotates against the direction of rotation of the workpiece.

*Surface grinding* is used to produce *flat surfaces.* The work may be ground by using either the *periphery* or the *end face* of a grinding wheel. *Form grinding* uses *specially shaped* grinding wheels to accurately finish surfaces previously machined to a special shape like *threads, gear teeth* and *spline shafts.* Complex forms like *punches* and *dies* can be ground by using special machines called *jig grinders .*

*Centreless grinding* is a method of grinding in which the workpieces are not held between centres (Fig 18.1). The work lies on a *work rest* between the wheels. Both grinding and regulating wheels rotate in the same direction. The work and the regulating wheel are fed forward, forcing the work against the grinding wheel. The axial movement of the work is obtained by tilting the regulating wheel.

### Abrasive wheels

The abrasives used in grinding wheels are small particles of *silicon carbide* or *aluminium oxide.* Grinding wheels are made by using a suitable material to *bond* the abrasive particles together. Different *particle sizes* and different kinds of *bonding materials* are used to make a whole range of grinding wheels, each suitable for a different type of work and finish. *Silicon carbide* wheels are used to grind low tensile strength materials, such as the tips of cutting tools, ceramics, cast iron, brass, etc. *Aluminium oxide* wheels are better suited for materials of higher tensile strength, such as most types of steel, wrought iron, tough bronzes, etc..

## Lapping

*Lapping* is an abrading process that is used to produce *geometrically true surfaces*, achieve *high dimensional accuracy*, secure a *fine surface finish*, and obtain a *close fit* between two contact surfaces. In the lapping process, fine particles of an abrasive material are mixed with oil to form a paste. Lapping is done by rubbing the surface of the work with an object called a lap in an *ever changing path*. The lap is made of a relatively *soft porous* material like cast iron or copper.The paste is rubbed into the lap, an action which causes the abrasive particles to become imbedded in it. Laps are operated by machine or by hand.

## Honing

*Honing* is usually used to finish internal or external cylindrical surfaces which have been previously machined or ground. The abrasive is in the form of a flat *stone* or *stick* called a *hone*. A few of these stones are mounted round a metal cylinder to form a *honing tool* which is reciprocated axially, while being in contact with the rotating work piece.

## Superfinishing

*Superfinishing* is a process which uses *large bonded abrasive stones* to produce a surface finish of extremely *high quality*. A reciprocating motion is given to a large stone which presses lightly on the work piece. The work piece itself is rotated or reciprocated depending on its shape.

## Polishing and buffing

*Polishing* is used to remove scratches and small defects from a surface. This is done by the cutting action of fine abrasive particles applied to wheels made of cloth, leather, felt, etc. The wheel is rotated while the workpiece is held against it.

**Buffing** is a refined kind of polishing in which a **mirror finish** (unobtainable by polishing) is produced. Buffing wheels are made of felt or sewed layer wheels of cloth. Very fine abrasives are applied to the wheel which is rotated at high speeds.

### Shot or grit blasting, shot peening, and hydrohoning

In **shot** or **grit blasting**, particles of abrasives or other materials moving at high velocity are made to strike the surface being treated. Scales, burrs, rust, etc. are removed and the surface acquires a matt appearance.

**Shot peening** is a process used to strengthen and harden a surface. In this process (which is similar to shot blasting), steel balls moving at high velocity strike the surface, which becomes work hardened and fatigue resistant.

**Hydrohoning** is used to smoothen surfaces by using a stream of liquid which carries abrasives. It is often used to remove burrs and marks in metal moulds.

### Barrel finishing

This process eliminates **hand finishing**, and is therefore very economical in the use of labour. The work pieces are placed in a many-sided barrel, which also contains abrasive materials (like stones, balls, abrasives, etc.) together with a suitable liquid. When the barrel is rotated for a suitable period of time, the **mutual impact** between the workpieces and the abrasive material removes **surface irregularities.**

---

## Vocabulary

| | | | |
|---|---|---|---|
| **abrasive** | Schleifmittel *n* | **hone** | honen *v* |
| **abrasive wheel** | Schleifscheibe *f* | **lap** | läppen *v* |
| **barrel finish** | trommelpolieren *v* | **periphery** | Rand *m* |
| **buff** | hochglanzpolieren, schwabbeln *v* | **polish** | polieren *v* |
| **bonding material** | Bindemittel *n* | **regulating wheel** | Regelscheibe *f* |
| **centreless grind** | spitzenlos schleifen *v* | **shot peen** | verfestigungs- strahlen *v* |
| **finishing process** | Endbearbeitungsverfahren *n* | **shot or grit blast** | körnchenblasen *v* |
| **grind** | schleifen *v* | **superfinish** | kurzhubhonen *v* |

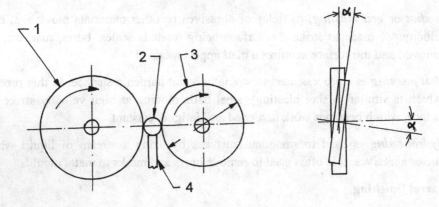

1. Grinding wheel, 2. Regulating wheel, 3. Work, 4. Work rest.

Fig 18.1 Principle of external centreless grinding

# Exercises XVIII

## 1. Answer the following questions:

a) How is metal removed from a metal surface by grinding ?

b) What kind of accuracy and surface finish does grinding produce ?

c) How much metal is removed from a surface by grinding ?

d) When is grinding the only method that can be used to remove metal from an object ?

e) State the names of the different grinding methods that are available.

f) What parts of a grinding wheel are used for surface grinding ?

g) What changes occur on the surface of an object when it is shot peened ?

h) What kind of surface finish is produced by a buffing process ?

i) What advantages does the  barrel finishing process have over other finishing processes ?

j) Give a brief description of what occurs in the shot blasting process.

## 2. Fill in the gaps in the following sentences :

a) Grinding is a _____ operation which gives high dimensional _____ to workpieces.

b) Very _____ metal is _____ from the surface in a grinding operation.

c) Surfaces which are too _____ to be machined by other _____ can be machined by grinding.

d) External cylindrical grinding is used to _____ cylindrical or _____ external surfaces.

e) Internal _____ grinding is used to produce cylindrical _____ or internal _____ on a workpiece.

f) In surface grinding, the _____ or the _____ of the grinding wheel may be used.

g) Buffing is a ____ kind of polishing in which a ____ is produced.

h) Shot ____ is a process used to ____ and ____ a surface.

i) Superfinishing is a process which uses ____ to produce a ____ of extremely high quality.

j) Barrel finishing eliminates ____ and is therefore very economical in the use of ____ .

## 3. Translate into English

a) Schleifen ist ein Verfahren, welches mit einer rotierenden Schleifscheibe durchgeführt wird und Metall von der Oberfläche eines Werkstückes abspant.

b) Schleifen ist normalerweise eine Endbearbeitung, welche eine hohe Oberflächengüte und eine sehr hohe Genauigkeit von Werkstücken erzeugt, die schon vorher durch andere Verfahren bearbeitet wurden.

c) Materialien, die zu hart sind, um durch andere Verfahren bearbeitet werden zu können, können ebenfalls durch Schleifen bearbeitet werden. Es wird nur eine kleine Menge von Metall beim Schleifprozess entfernt.

d) Außen-Rund-Schleifen wird dazu eingesetzt, zylindrische oder kegelige Oberflächen auf der Außenseite des Werkstückes herzustellen. Das Oberflächenschleifen wird eingesetzt, um flache Oberflächen mit der Fläche oder dem Rand der Schleifscheibe herzustellen.

e) Die Schleifmittel, die bei Schleifscheiben zum Einsatz kommen, sind kleine Körner aus Siliziumkarbid oder Aluminiumoxyd. Schleifscheiben werden mit einem passenden Material, welches die Schleifpartikel zusammenklebt, hergestellt.

# 19 The manufacture of plastic goods

Plastic manufacturing processes are very *cost-effective*, because products having complicated shapes can be *moulded* in a *single operation,* and no further work needs to be done on them. However the moulds used are very expensive, and such processes can only be profitable if large quantities of goods are produced. The raw materials used in the manufacture of plastic goods are usually in the form of *powder, granules,* or *liquid.*

The most commonly used plastic materials can be classified into two types, *thermoplastics* which can be moulded several times, and *thermosetting plastics* which can be moulded only once (see chapter 1).

## 1. The injection moulding process

This is probably the most widely used process for the manufacture of moulded thermoplastic goods. Granules of the raw plastic material are fed into a heated *plasticizing cylinder* through a *screw feed mechanism* (Fig 19.1). The plastic material is heated, compressed, and degassed until it is in a *soft state*.

The screw feed mechanism is now given a sudden *push forward*. This forces the soft plastic through an injection nozzle into a *two piece mould*. The mould is cooled rapidly, causing the plastic in the mould to harden quickly. After the plastic has *hardened*, the mould opens and the finished component is finally *ejected* from the mould.

## 2. The extrusion process

One of the commonest ways of producing *bar, tube, sheet, film etc.* from thermoplastic materials is *by extrusion*. The machines used have a screw feed mechanism, similar to that in the injection moulding process. The screw feed mechanism forces the soft plastic through an *opening* in a *die* which has the desired *cross-section*. The soft plastic comes out of the opening *in the form of a strand* and is hardened by cooling in a stream of air. *Semifabricated products* like sheets, bars,etc. produced by extrusion, can be used for the fabrication of more complex plastic products.

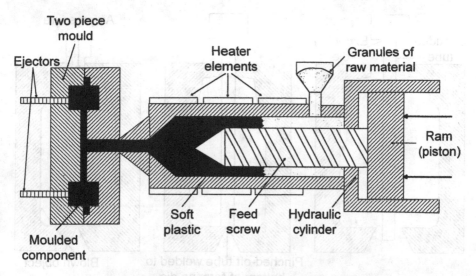

Fig 19.1 Diagram of an injection moulding machine

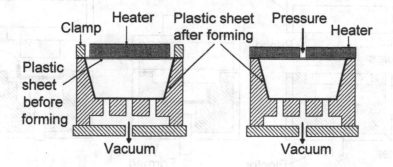

Fig 19.2 Vacuum forming

Fig 19.3 Pressure assisted
vacuum forming

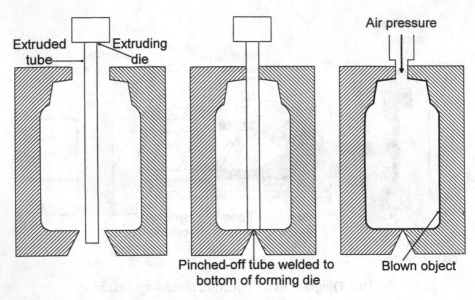

Fig 19.4 Extrusion blow moulding

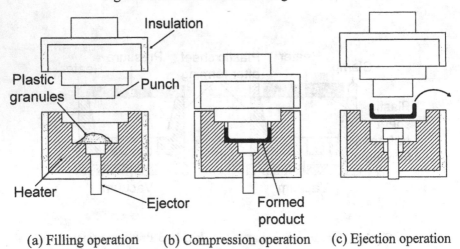

(a) Filling operation    (b) Compression operation    (c) Ejection operation

Fig 19.5 Compression moulding

# Exercises XIX

## 1. Answer the following questions:

a) What are the names of the two most commonly used types of plastic material ?

b) Why are the processes used in the manufacture of plastic goods extremely cost-effective ?

c) Why are these manufacturing processes unsuitable for small batch production ?

d) Which is probably the most widely used process for the manufacture of thermoplastic goods ?

e) How are semifabricated products like tubes, sheets etc. produced from raw plastic materials ?

f) What type of mechanism is used to feed raw plastic granules into the plasticizing cylinder in an injection moulding process ?

g) In what way is the pressure assisted vacuum forming process different from the ordinary vacuum forming process ?

h) In what way are goods made from thermosetting plastics different from those made from thermoplastics ?

i) What types of objects are made by thermoforming proceses ?

j) What is done in the extrusion process to ensure that the plastic product hardens quickly ?

## 2. Fill in the gaps in the following sentences:

a) The _____ used in the manufacture of plastic goods are in the form of _____ .

b) Plastic manufacturing processes are _____ because complicated shapes can be _____ in a single operation.

c) Granules of the _____ are fed into a _____ through a screw feed mechanism.

d) The screw feed mechanism is pushed to force the ____ through an ____ into a two-piece mould.

e) The plastic is ____, compressed, and degassed, until it is in a ____.

f) In vacuum forming the sheet is ____ and a vacuum is ____ from below.

g) The ____ is cooled rapidly, and the plastic object hardens ____.

h) In the blow moulding process, pieces of extruded soft plastic are ____ and welded ____.

i) The moulded object is removed ____ by separating the two ____.

j) Objects made from ____ plastics are ____ hardened.

## 3. Translate into English:

a) Kunststoffherstellungsverfahren sind sehr kostengünstig, da komplizierte Formen in einem einzigen Prozess geformt werden können. Jedoch sind die Gießformen sehr teuer und solche Prozesse sind nur gewinnträchtig, wenn große Mengen produziert werden.

b) Dieses ist wahrscheinlich der am häufigsten eingesetzte Prozess für die Herstellung gegossener Kunststoffprodukte. Körner des Rohkunststoffmaterials werden über eine Schnecke einem beheizten Plastifizierzylinder zugeführt.

c) Einer der gebräuchlichsten Wege Stäbe, Röhren, Folien usw. aus thermoplastischem Material herzustellen, ist das Strangpressen. Die Schneckenstrangpresse drückt das weiche Material durch die Öffnung in ein Schneidwerkzeug, welches den gewünschen Querschnitt besitzt.

d) Beim Vakuumformen wird die Scheibe im Umfang eingespannt und erhitzt. Ein Vakuum wird von unterhalb eingesetzt, um die Scheibe in ein weibliches Formeisen zu ziehen. Nach dem Abkühlen nimmt die Kunststoffscheibe die Form des Formeisens an.

# 20 Numerically controlled machines

*Automatic lathes* have been in use for a long time, and are mostly used for the rapid production of *large numbers* of *identical components*. A more recent development has been the use of *numerically controlled machines* for the *production* of *small batches* of components. These machines need not be reset each time a new component has to be produced.

## Automatic lathes

*Single* and *multispindle* automatic lathes have been found to be most useful for the manufacture of *large numbers* of *identical components*. The material used is usually in the form of metal rods, which are fed in through the hollow spindles of the lathes. Mechanical devices like *cams* and *stops* are used to carry out a series of operations according to a *predetermined program.* The setting-up time is usually long, and such lathes are usually *unsuitable* for *small batch* production.

## Computer Numerical Control

Most components required by industry need to be produced only in *small* and *medium sized batches*, and their production does not justify the use of automatic lathes. In the past, the production of small and medium sized batches was achieved by the use of *conventional machine tools* and *highly skilled labour*. This resulted in a *low machine utilization time*, a *long waiting time*, and a *high cost of production*. The accurate production of components according to a definite time schedule was extremely difficult. A solution to this problem was found by the development of a system of automatic production. In this system, the machine produces a component automatically in accordance with predetermined instructions, which can be easily changed when another component is to be manufactured. The use of a programmed computer in machine tools has led to this type of automation being called *computer numerical control (CNC).*

In computer numerical control, the operation of a machine tool is controlled by a program run on a computer. The machine tools used in this type of work have *movements* and *drives* which are *different* from those in *conventional* machine tools. They have *electrical drives* which are *controlled by a computer*. The machines do not need *to* be set up by *a skilled operator,* each time a new type of component is made. They work automatically under computer control, and only a change in the *computer program* is required to produce a new batch of components. Such a system has the following advantages:

1. The flexibility of the system results in high machine utilization times.

2. Deliveries can be made on time because production is more predictable.

3. Special tooling (e.g using jigs or fixtures) is not required.

4. Consistency in the quality of the goods produced reduces inspection times.

5. Non-machining time is reduced due to automatic tool changing, automatic clamping, etc.

6. Design changes can be easily made, because only the program has to be changed to produce a new component.

7. The skill of the operator is unimportant, because the accuracy is only dependent on the machine and the program.

8. Estimation of costs is easier than in previous methods of production.

**Machining and turning centres**

Two major types of CNC machine tools are *machining centres* and *turning centres.* A machining centre is a machine which can perform milling, drilling, boring, reaming, and tapping operations, and is similar to a *milling machine.* Turning centres are basically similar to *lathes.* In addition to the usual, shaft, bar, and chuck work, operations like contour milling, tapping, making T-slots and keyways, etc. can be done by turning centres. These machines have many features which ensure accurate performance and productivity over long periods of time. Some of the features are the following :

1. The machines are of massive construction. This gives them the stability needed to withstand large cutting forces, and the thermal effects caused by the large quantities of chips produced.

2. Minimization of wear is achieved by the use of rolling components instead of sliding components.

3. The screw threads of conventional machines are replaced by circulating ball screws. Rigid mounting techniques for bearings and screws and also preloading are techniques used for overcoming backlash.

4. Each tool is held in an adaptor, and stored in a drum, chain, or other type of magazine. The correct tool for an operation is selected automatically from a magazine, and is used to replace the one already existing on the machine.

5. Measuring instruments give a continuous indication of the position of the cutting tool. These are compared with the desired values by the computer and corrected when necessary. The use of such a technique automatically compensates for tool wear.

---

# Vocabulary

| | | | |
|---|---|---|---|
| accordance | Übereinstimmung *f* | hollow | hohl *adj* |
| achieve | leisten *v* | identical | identisch *adj* |
| adaptor | Aufnehmer *m* | important | wichtig *adj* |
| automatic | automatisch *adj* | indication | Anzeige *f* |
| backlash | Flankenspiel *n* | instruction | Befehl *m* |
| batch | Menge, Losgröße *f* | machining centre | Bearbeitungszentrum *n* |
| compensate | kompensieren *v* | numerical control | Numerische Steuerung *f* |
| construction | Anlage *f* | predetermine | vorherbestimmen *v* |
| continuous | ständig *adj* | predictable | voraussagbar *adj* |
| conventional | herkömmlich *adj* | previous | vorher *adj* |
| development | Entwicklung *f* | recirculating | umlaufend *adj* |
| difficult | schwer *adj* | select | auswählen *v* |
| drum | Trommel *f* | setting-up | Einstellung *f* |
| estimation | Schätzung *f* | sliding | gleitend *adj* |
| exist | existieren *v* | stability | Stabilität *f* |
| extremely | sehr *adv* | turning centre | CNC Drehmaschine *f* |
| feature | Eigenschaft *f* | utilization | Nutzung *f* |
| flexible | beweglich *adj* | | |

# Exercises XX

## 1. Answer the following questions :

a)  What are automatic lathes used for ?

b)  How were small batches of components produced in the past ?

c)  What were the disadvantages of the previous methods of production ?

d)  How is nonmachining time reduced in a CNC machine ?

e)  In what way do the drives in CNC machines differ from those in conventional machines ?

f)  Why does the use of CNC result in high machine utilization times ?

g)  What are the changes that have to be made, when a new component has to be produced using a CNC machine ?

h)  Why is the skill of the operator unimportant in a CNC machine ?

i)  How are tools changed in a CNC machine ?

j)  How is the wear on a CNC machine minimized ?

## 2. Fill in the gaps in the following sentences :

a)  Automatic lathes are ____ for the ____ of a large number of ____ components.

b)  A new ____ has been the use of ____ controlled machines for ____ production.

c)  Previously, small batch production was ____ using ____ labour and ___ machine tools.

d)  This resulted in a low machine ____ time and a high ____ of production.

e)  The ____ of a CNC machine is ____ by a computer program.

f)  Special tools like jigs and ____ are not ____ when a CNC machine is used.

g) Minimization of wear in CNC machines is achieved, by using ____ components instead of ____ components.

h) Measuring ____ give a ____ indication of the position of the ____ tool.

i) The ____ tool for a particular operation is ____ automatically from a ____.

j) The ____ construction of the machines gives them the ____ needed to ____ large cutting forces.

## 3. Translate into English:

a) Einfach- und Mehrspindeldrehautomaten haben dort ihre größte Anwendung gefunden, wo eine große Anzahl von identischen Werkstücken hergestellt werden muß. Eine neuere Entwicklung ist der Gebrauch von computergesteuerten Maschinen für die Produktion von kleinen Losgrößen.

b) In der Vergangenheit wurde die Produktion von kleinen und großen Mengen durch den Gebrauch von konventionellen Werkzeugmaschinen geleistet und zudem durch gut ausgebildete Arbeitskräfte. Hieraus resultiert eine niedrige Maschinennutzungszeit, eine lange Wartezeit und hohe Produktionskosten.

c) Die Tätigkeit der CNC-Werkzeugmaschine wird durch einen Computer gesteuert. Die Werkzeugmaschinen, die bei einer solchen Art der Arbeit benutzt werden, weisen nicht die normalen Bewegungen und Antriebe auf, die bei konventionellen Maschinen gebraucht werden.

d) CNC-Maschinen müssen nicht durch einen hoch ausgebildeten Arbeiter eingerichtet werden. Nur eine Änderung im Computerprogramm ist erforderlich, um ein neues Los von Teilen herzustellen. Aus der Flexibilität des Systems resultiert eine hohe Maschinennutzungszeit.

# 21   The automobile engine

## Internal combustion engines

*Internal combustion engines* are engines (or motors) in which the fuel is burnt inside the engine itself, as compared with *external combustion engines* like the *steam engine* where the fuel is burnt outside the working cylinder of the engine. The *petrol engine* and the *diesel engine* as used in modern automobiles are both internal combustion engines. The *chemical energy* of the fuel is converted into *heat energy* by combustion, and part of this heat energy is converted into *mechanical energy* by the engine.

## The petrol engine

Most petrol engines used today are based on the *Otto four stroke cycle* in which a mixture of fuel (in vapour or gaseous form) and air is *compressed* and then *ignited.* Fig 21.1 shows the principle of operation of a four stroke single cylinder petrol engine. It consists of a *cylinder* in which a *tight-fitting piston* can undergo an up and down (or reciprocating) motion. At the top or *head* of the cylinder are two valves, the *inlet valve* and the *outlet valve.* These valves can be *opened* or *closed* to allow a *gas-air mixture* to enter or leave the cylinder. When both valves are closed, the cylinder becomes a *gas-tight chamber,* and any upward movement of the piston *compresses* the *gas-air mixture* contained in the cylinder.

The piston is connected through a *connecting rod* to a *crankshaft.* This mechanism enables the *reciprocating motion* of the piston to be converted into the *rotatory motion* of the crankshaft (see Chap 10).

## The four stroke cycle

In a four stroke cycle, the movement of the piston takes place in four stages, two upwards and two downwards, each of these stages being called a *stroke.* Only one in every four strokes is an *ignition* or *power stroke*, in which the engine *extracts energy* from the fuel. The power stroke takes place once in every two revolutions of the crankshaft. The four strokes of the piston are shown diagramatically in Fig 21.1 and are called, *induction, compression, power* (or ignition), and *exhaust.*

1. During the *induction stroke*, the inlet valve opens and the piston moves downwards. This movement creates a *partial vacuum* that sucks in a petrol-air mixture (which comes from the carburettor) through the open inlet valve into the cylinder. The exhaust valve remains closed during this stroke.

2. During the *compression stroke*, both inlet and exhaust valves remain closed. The piston moves upwards compressing the petrol-air mixture.

3. At the beginning of the ignition or *power stroke*, the spark plug produces a spark which ignites the compressed mixture. The large amount of heat generated during the combustion process increases the pressure of the burnt gases in the cylinder, and the force resulting from this increased pressure pushes the piston downwards.

4. During the *exhaust stroke*, the exhaust valve opens and the piston moves upwards pushing the burnt gases out of the cylinder. At the end of this stroke, the whole cycle is repeated indefinitely until the engine is stopped.

### Crankshafts and camshafts

The reciprocating motion of the piston causes the connecting rod to move, and this in turn gives the crankshaft a rotatory motion. Mounted at the end of the crankshaft are two wheels. One is a heavy wheel called a *flywheel* which *stores* the *energy* generated during the power stroke, and uses this stored mechanical energy to move the crankshaft, connecting rods, pistons, etc. through the three other *idle strokes*. The other wheel with teeth on it called a *sprocket wheel*. The sprocket wheel drives the *camshaft* through a chain drive. The camshaft has the function of *opening* and *closing* the valves at the right time.

Fig 21.2 and Fig 21.3 show two arrangements for driving valves using a camshaft. Fig 21.4 shows the crankshaft and associated components of a four cylinder engine, while Fig 21.5 shows a camshaft.

### Multicylinder engines

A single cylinder engine has poor *engine balance*, and needs a very *heavy flywheel*. Much smoother operation is achieved by using *multicylinder engines*, which have more power strokes per crankshaft revolution. The four cylinder engine is the most popular, as it achieves a good compromise between price and performance. Six and eight cylinder engines run more *smoothly* and *silently*, but are more expensive to build.

## The Diesel or compression-ignition engine

The *diesel engine* resembles the petrol engine as far as its *mechanical components* like the cylinder, piston, crankshaft, etc. are concerned. It has also four strokes (which are similar to those in the petrol engine), these being induction, compression, power, and exhaust, with only the power stroke producing power.

The *chief difference* between the two engines, lies in the *different methods* used for *introducing fuel* into the cylinder, and for *burning fuel* in the cylinder. In the diesel engine:

1. *Spark plugs* are not required.

2. *Only air* is sucked in during the induction stroke, as compared with the petrol-air mixture in a petrol engine.

3. The *carburettor* is replaced by a *fuel pump*, which sprays a small quantity of fuel directly into the cylinder, at the end of the compression stroke. The (adiabatic) compression of the air in the cylinder during the compression stroke, raises the *air temperature* to a high enough value to cause the fuel to ignite when it is injected at the end of the compression stroke. Diesel engines are also known as *compression-ignition* engines.

Diesel engines have a *higher thermal efficiency* than petrol engines, and therefore consume less fuel. However, they are noisier, and less smooth in operation than petrol engines.

## Recent improvements in internal combustion engines

Manufacturers are always trying hard to reduce the fuel consumption of internal combustion engines and to improve their efficiency. At the same time efforts are being made to reduce the emission of unwanted gases from the exhausts of engines. Among the improvements made in recent years are the following:

1. The use of four valves instead of two valves per cylinder.

2. The use of a fuel injection pump in petrol engines instead of a carburettor.

3. The use of higher compression ratios with appropriate high octane fuels.

4. The use of five speed instead of four speed gear boxes.

5. The use of light alloys instead of cast iron and steel in engine construction.

6. The use of catalysers to reduce unwanted exhaust emissions.

Work is also being done on the use of other sources of fuel for engines, like methane gas, biodiesel, fuel cells, solar cells, etc.

---

## Vocabulary

| | | | |
|---|---|---|---|
| **associated** | verbunden *adj* | **ignite** | entzünden *v* |
| **balance** | Gleichgewicht *n* | **inlet valve** | Einlassventil *n* |
| **based on** | begründet auf *v* | **induction** | Einführung *f* |
| **burn** | verbrennen *v* | **mixture** | Mischung *f* |
| **camshaft** | Nockenwelle *f* | **noise** | Geräusch *n* |
| **chain** | Kette *f* | **outlet valve** | Auslassventil *n* |
| **chamber** | Kammer *f* | **petrol** | Benzin *n* |
| **combustion** | Verbrennung *f* | **popular** | beliebt *adj* |
| **compress** | zusammendrücken *v* | **power** | Kraft *f* |
| **concerned** | betroffen *adj* | **push** | schieben *v* |
| **connecting rod** | Pleuelstange *f* | **repeat** | wiederholen *v* |
| **convert** | umwandeln *v* | **revolution** | Umdrehung *f* |
| **crankshaft** | Kurbelwelle *f* | **smooth** | zügig *adj* |
| **create** | erzeugen *v* | **spark plug** | Zündkerze *f* |
| **efficiency** | Leistungsfähigkeit *f* | **sprocket** | Kettenrad *n* |
| **exhaust** | Auspuff *m* | **stroke** | Hub *m* |
| **extract** | ausscheiden *v* | **suck** | saugen *v* |
| **flywheel** | Schwungrad *n* | **tight** | dicht *adj* |
| **fuel** | Brennstoff *m* | **vacuum** | Vakuum *n* |
| **generate** | erzeugen *v* | **vapour** | Dampf *m* |

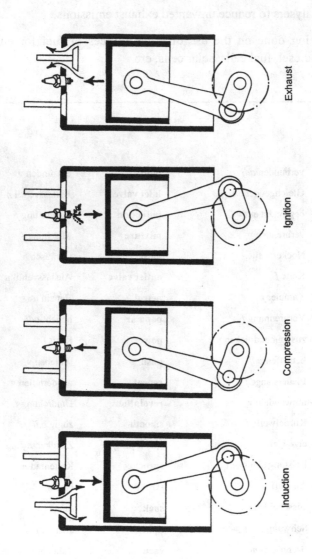

Fig 21.1 Four stroke cycle

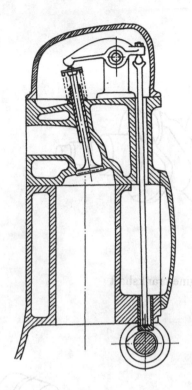

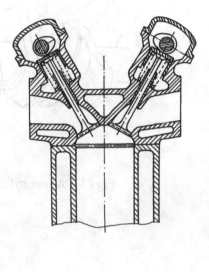

Fig 21.2 Pushrod overhead
valve layout

Fig 21.3 Twin overhead
camshaft layout

Fig 21.4 A four cylinder engine crankshaft

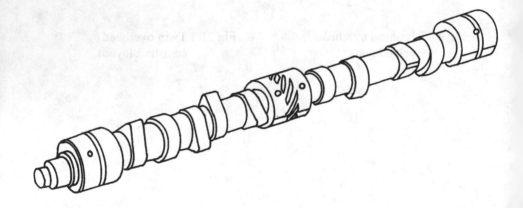

Fig 21.5 A four cylinder engine camshaft

# Exercises XXI

## 1. Answer the following questions:

a) What is the difference between an internal and an external combustion engine ?

b) What energy conversion processes take place in an internal combustion engine ?

c) How many strokes has an Otto cycle petrol engine ?

d) What are the functions of the connecting rod and the crankshaft ?

e) In what way is the power stroke different from the other strokes ?

f) What are the four strokes in the Otto cycle called ?

g) What happens during the induction stroke ?

h) What is the purpose of the flywheel ?

i) What are the main differences between a petrol and a diesel engine ?

j) What advantages do multicylinder engines have over single cylinder engines ?

## 2. Fill in the gaps in the following sentences :

a) In an ____ combustion engine, the ____ is burnt inside the ____

b) The ____ energy of the fuel is ____ into heat energy by combustion, and part of this is converted into ____ energy by the engine.

c) The ____ can be opened or ____ to allow a ____ mixture to enter or leave the cylinder.

d) The connecting rod and the ____ enable the ____ motion of the piston to be converted into ____ motion.

e) During the ____ stroke, the engine ____ energy from the fuel.

f) During the induction stroke, the ____ valve opens and the ____ moves ____.

g) During the ____ stroke both valves remain closed and the piston moves upwards ____ the petrol-air mixture.

h) The ____ wheel drives the ____ through a chain drive.

i) The difference between a petrol and a ____ engine lies in the different methods used for ____ and ____ fuel in the cylinder.

j) Diesel engines have a higher ____ than petrol engines and therefore ____ less fuel.

## 3. Translate into English:

a) Der Benzinmotor und der Dieselmotor sind beide Verbrennungsmotoren, in welchen Kraftstoff innerhalb des Motors verbrannt wird. Die chemische Energie des Kraftstoffes wird zuerst durch Verbrennung in Wärmeenergie und dann durch den Motor in mechanische Energie umgewandelt.

b) Die meisten Benzinmotoren basieren auf dem Otto-Viertakt-Prinzip, bei dem ein Gemisch aus Benzin und Luft zuerst komprimiert und dann gezündet wird.

c) Der Benzinmotor hat einen Zylinder mit einem starr-geführten Kolben, welcher eine Auf- und Abbewegung durchführen kann. Am Kopf des Zylinders befinden sich das Einlassventil und das Auslassventil. Diese Ventile können geöffnet oder geschlossen werden, um einem Gas-Luftgemisch zu ermöglichen, in den Zylinder einzuströmen oder den Zylinder zu verlassen.

d) Bei einem Viertakt-Prozess findet die Bewegung des Kolbens in vier Schritten statt, zwei Aufwärts- und zwei Abwärtsbewegungen. Diese vier Hübe werden Ansaugen, Verdichten, Verbrennen und Ausströmen genannt. Nur während des Verbrennungshubes wird Energie vom Kraftstoff genutzt.

e) Der Benzinmotor und der Dieselmotor sind ähnlich, soweit es die mechanischen Komponenten betrifft. Der hauptsächliche Unterschied zwischen den Maschinen liegt in den verschiedenen Methoden, die für das Einströmen des Kraftstoffes in den Zylinder und für das Verbrennen des Kraftstoffes angewendet werden.

# 22 The manufacturing enterprise

The manufacturing enterprise is a complex organization consisting of a number of departments whose activities are closely interrelated. In most enterprises, the different activities are usually coordinated and arranged in a *closed loop* as shown in Fig.22.1. The figure also illustrates the important role played by *computers* and *common data banks* in the activities of manufacturing enterprises. Some of the most important activities are discussed below.

### 1. Sales, market research, and market forecast

Before a new product is manufactured, suitable markets have to be identified, their magnitudes assessed, and the extent of the present and future competition appraised. *Product preferences* may differ from country to country, or from age group to age group. These and other factors affecting sales have to be carefully studied. *Market studies* of various types are commonly used to identify the kinds of products that can be profitably manufactured and sold in a given environment.

### 2. General product concept and design

The initial design of a product is usually done by a number of *design specialists*. Industrial designers, mechanical and electrical engineers, materials science and other specialists, work in close cooperation at this stage.

A well designed product must be *visually appealing*, and have *the ability* to *perform its function* satisfactorily and reliably through its expected life. It should be easy to use, easy to maintain, and easy to manufacture at a *competitive price*. It should not place undue *physical strain* on the user. A further constraint imposed on manufactured products is an *environmental* one. Products should be made in such a way that they can be easily disposed of or recycled at the end of their useful life without causing damage to the environment.

### 3. Detailed product design

Once the general design of the product has been finalized, it is necessary to produce *detailed drawings* of the various components and of the assembled product. Computer aided design (CAD) is nearly always used today, and the drawings are stored in a *common data base* which is used as a basis for computer aided manufacturing (CAM).

In some cases it may be *cheaper* and more convenient to purchase standard *mass produced components* than to manufacture them. It may also be advantageous to purchase large *subassemblies* or specialized components from outside suppliers. Such a procedure is called *outsourcing*.

When the detailed design process has been finished, a *bill of materials* (i.e, a list of all required materials, components, assemblies, etc.) is prepared. This is a document that is *central* to the whole *manufacturing process*.

### 4. Production planning

In the production planning stage, the most appropriate *materials, processes, and process parameters* have to be chosen.This is no easy task, given the large number of materials and processes which are available today.

*Tooling* has to chosen, and suitable *dies, jigs, and fixtures* have to be made. *Fixtures* are devices which hold a workpiece in place while work is being done on it. *Jigs* are devices which hold the workpiece in place and also *guide* the *tools* which do the machining.

*Group technology* is used to identify and manufacture parts which have similar features. It is more economical to use group technology and *manufacture families of parts* with similar features using the same sequence of processes, than it is to manufacture them individually.

### 5. Process research and development

*Research* and *development* have to play an important role in the activities of a manufacturing enterprise if it is to remain competitive. Old processes have to be improved, and new processes have to be developed.

*Computer modelling* and *process simulation* are techniques which are widely used to optimize manufacturing processes. It is also necessary to consider the impact of *manufacturing* on the *environment*. Noise, gaseous fumes, smoke, toxic wastes, water pollution, etc. should be minimized, and international environmental standards should be maintained.

## 6. Manufacturing

In manufacturing, the layout of machines and other equipment should be optimized to facilitate the flow of raw materials, tools, and components in different stages of production.The *quality* and *dimensions* of components need to be systematically checked and *corrective action* taken where necessary.

Special equipment may be needed to handle, move, and store finished components as well as tools, jigs, and fixtures. The final stage consists of the *assembling* of the components into the *final product*, which is then ready for delivery.

## 7. Production control and material requirements planning (MRP)

Efficient production is a difficult task, and stringent control is necessary if competitive advantages are to be maintained.

a) Programs for controlling *product quality* and for the *maintenance* of *machines* and **tools** are essential in any manufacturing activity

b) Careful monitoring of *machine utilization* and *labour performance* is necessary if efficiency in production is to be maintained.

c) *Material requirements planning (MRP)* is the name given to a modern and comprehensive approach to the *planning, scheduling,* and *controlling of production*. The number and types of units produced can vary from day to day depending on customer orders, market forecasts, and production plans. Material requirements planning is driven by *consumer demand*, and is clearly *different* from the *technical planning* of production. It is however equally important, and is discussed in detail in the next chapter.

## 8. Shipping and delivery

The manufactured products have to be delivered to *wholesalers, retailers*, and *individual customers*. Some of the manufactured products and also products from outside suppliers may be stored, and are then part of the inventory. The *inventory control* department feeds back sales information to the production department, so that production can be adjusted to meet *fluctuating demands* for end products.

## 9. Customer service

Products delivered to the customer need to be *serviced* and *repaired* during their useful lifetime. *Information* and *statistics* concerning product faults and weaknesses, product reliability, customer satisfaction, etc. needs to be collected and sent to the market research and production departments. This *feedback* of *information* closes the loop which coordinates the activities of an enterprise as shown in Fig.22.1. Such feedback is necessary for the improvement of old products, for the development of new products, and for improving the efficiency of the manufacturing process.

## 10. The recycling and disposal of products

Environmental and energy considerations have led to the introduction of stringent measures for the disposal and recycling of products at the end of their useful life. Products must be disposed of in ways that are both *economically* and *ecologically* acceptable. *Recycling* enables the materials in the products to be reused.

---

# Vocabulary

| | | | |
|---|---|---|---|
| appeal | appellieren v | enterprise | Unternehmen n |
| appraise | schätzen v | environment | Umgebung f |
| assemble | montieren v | feedback | Rückkopplung f |
| assess | bewerten v | forecast | voraussagen v |
| closed loop | geschlossener Kreis m | impose | aufdrängen v |
| comprehensive | umfassend adj | inventory | Bestandsverzeichnis n |
| consist of | bestehen aus v | layout | Anordnung f |
| constraint | Zwang m, Nötigung f | magnitude | Größenordnung, Größe f |
| delivery | Lieferung f | material requirements | deterministische |
| | | planning | Bedarfsmittlung f |
| demand | Bedarf m | preference | Bevorzugung f |
| department | Abteilung f | recycle | wiederbewerten v |
| design | entwerfen v | schedule | Arbeitsterminplan m |
| dispose of | beseitigen, loswerden v | service | Dienstleitung f |

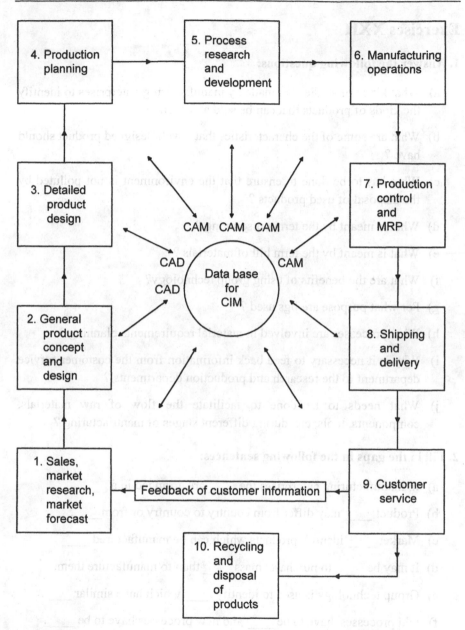

Fig.22.1 Diagram showing the coordination of the activities of a
manufacturing enterprise

# Exercises XXII

## 1. Answer the following questions:

a) What kinds of studies are made by manufacturing enterprises to identify the kinds of products that can be sold at a profit ?

b) What are some of the characteristics that a well designed product should have ?

c) What has to be done to ensure that the environment is not polluted by the disposal of used products ?

d) What is meant by the term outsourcing ?

e) What is meant by the term bill of materials ?

f) What are the benefits of using group technology?

g) For what purpose are jigs used ?

h) What processes are involved in material requirements planning ?

i) Why is it necessary to feed back information from the customer service department to the research and production departments ?

j) What needs to be done to facilitate the flow of raw materials, components, tools, etc. during different stages of manufacturing ?

## 2. Fill in the gaps in the following sentences:

a) In manufacturing enterprises the _____ are arranged in a _____ .

b) Product _____ may differ from country to country or from_____.

c) Market _____ identify products which can be manufactured _____.

d) It may be _____ to purchase mass _____ than to manufacture them.

e) Group technology is used to identify _____ which have similar _____.

f) Old processes have to be _____ and new processes have to be _____.

g) The quality and _____ of the components have to be _____ checked.

h) The final stage consists of the ____ of the components into the ____ .

i) Programs for ____ product quality and for the ____ of machines are essential.

j) The ____ of customer information closes the loop which ____ the activities of a manufacturing enterprise.

## 3.Translate into English

a) Verschiedene Typen von Marktstudien werden eingesetzt, um die Art des Produkts zu identifizieren, das in einer bestimmten Umgebung gewinnträchtig hergestellt und verkauft werden kann.

b) Ein gut ausgelegtes Produkt muss ansprechend aussehen und die Fähigkeit besitzen, seine Funktion während der erwarteten Lebensdauer zufriedenstellend zu erfüllen.

c) Eine weitere Einschränkung die hergestellten Produkten auferlegt wird, ist umweltbezogen. Produkte sollten so hergestellt werden, dass sie am Ende der normalen Lebensdauer einfach entsorgt oder wiederverwertet werden können.

d) Im Schritt der Fertigungsplannung müssen die geeignetsten Materialien,Verfahren und Verfahrensparameter ausgewählt werden. Dieses ist keine leichte Aufgabe, betrachtet man die Vielzahl der heute verfügbaren Materialien und Verfahren.

## 23 The computer control of manufacturing systems

The traditional approach to manufacturing has been to view it as a *sequence* of *individual steps*. Competitive pressures have however exposed the short-comings of this view and compelled modern manufacturing enterprises to adopt a *system approach*, in which the various manufacturing activities form part of an *interacting dynamic system*.

The control and coordination of complex manufacturing systems has been greatly facilitated by the availability of cheap and fast computers. Almost every activity of a manufacturing enterprise from *product design* to *accounting* and *invoicing*, is controlled today by a hierarchy of computers.

**Computer aided design (CAD)**

Detailed product design was done in the past by making drawings on a *drawing board*. The modern method which is called computer aided design (CAD) uses a computer with suitable software. The object which is being designed can be viewed on a video display terminal, and the design data can be stored in a *common* **data base** (see Fig. 22.1). By using *geometric modelling and analysis*, different design variations can be studied as required. A particular advantage of this method is that the designed object can be *rotated* and *viewed* from *different angles* on the display screen. Some of the advantages of CAD are the following:

- Designs can be optimized in a short time, and design changes can be made rapidly in response to customer demand.

- Components and assemblies can be designed with the certainty that they fit together.

- Libraries of standard components can be compiled.

- Bills of materials can be prepared and drawings can be produced on paper using a graphical plotter. All design data can be stored in a common data base which provides the basis for later computer aided manufacturing.

## Computer aided manufacturing (CAM)

Computers are used to control many aspects of manufacturing, and this field of application has been called computer aided manufacturing (CAM). Some of the advantages of CAM are stated below.

- Numerical control (NC) and Computer numerical control (CNC) are used to control the machines which produce the components.

- The design data generated in the CAD process already exists in the common data base, and can be used to *program* the machines used in the manufacturing process.

- Computers are also used for *many other tasks* like process control and optimization, materials management, material movement, control of transfer lines and robots, monitoring, scheduling, reporting, etc.

## The integration of CAD and CAM into a common (CAD/CAM) process

An increase in efficiency can be achieved by integrating CAD and CAM into a common *CAD/CAM process*. A common data base is necessary to enable the entire process from the design stage to the manufacturing stage to be automated. A *two way flow of information* between design and manufacturing leads to a critical assesment of the capabilities and limitations of both materials and manufacturing processes. The use of CAD/CAM allows a *continuous review* and *improvement* of existing design, manufacturing, and production control processes.

## Material requirements planning (MRP)

Material requirements planning which is different from technical planning, is used for the *management of inventories* and the *timing of the requirements* of materials, components, subassemblies, etc. It consists of the following steps.

1. *A program of production* is first established based on the actual and anticipated demand. This must contain information about *batch sizes* and *delivery dates* for all *end products*.

2. *Lists of all small components and subassemblies* (which are required in the manufacture of the end products) are compiled using bills of materials files.

3. These requirements are compared with the *inventory on hand* (i.e, items in stock or being produced) to determine the *net requirements* for purchasing or manufacturing.

4. The next stage is *releasing,* which is the process of *initiating production orders*. Here, all information required for manufacturing like material lists, process routing, and performance tracking must be included.

5. This is followed by *scheduling,* which is the process of *assigning each order* to a *sequence of machines* taking into account batch sizes, delivery time, production capacity, and lead times. (Lead time is the time taken between the placing of an order and the delivery of the product.)

6. The final stage is *reporting,* which is the process of *informing management* about how the manufacturing process is proceeding in relation to schedules. This must include performance reports, deviations and cancellations from schedules, inventory status, etc.

**Computer integrated manufacturing (CIM)**

There has been an increasing trend towards automatizing the entire manufacturing process from design to finished product. A lot of progress has made in this direction, but *complete automation* of all aspects of manufacturing *has still to be achieved.*

**Limitations of computer control in manufacturing**

The proper use of computers facilitates the manufacturing process, and increases the competitiveness of an enterprise. However the use of computers alone *does not* guarantee *success*. Designs and processes which are *ill-conceived, badly planned* or *outdated*, cannot be improved significantly by the use of computer control. It is essential that the *physical principles* involved be clearly understood, materials, processes, and process parameters be correctly chosen, and that a *careful review* be carried out to *detect flaws* and *weaknesses* in the manufacturing system. A control system using computers or other devices merely executes the plans which are essentially made by human beings.

# Vocabulary

| | | | |
|---|---|---|---|
| **accounting** | Buchführung *f* | **performance** | Leistung *f* |
| **approach** | Methode *f*, Zugang *m* | **pollution** | Umweltverschmutzung *f* |
| **appropriate** | passend, geeignet *adj* | **prepare** | vorbereiten *v* |
| **available** | vorhanden, erhältlich *adj* | **progress** | Fortschritt *m* |
| **check** | prüfen *v* | **provide** | besorgen v |
| **compel** | zwingen, nötigen *v* | **purchase** | kaufen *v* |
| **competition** | Konkurrenz *f* | **release** | Auftrag erteilen *v* |
| **conceive** | vorstellen *v* | **report** | Bericht erstatten *v* |
| **coordinate** | richtig anordnen *v* | **reuse** | wiederverwenden *v* |
| **enterprise** | Unternehmen *n* | **route** | weiterleiten *v* |
| **execute** | durchführen *v* | **schedule** | Arbeitsterminplan *m* |
| **extent** | Bereich *m* | **sequence** | Reihenfolge *f* |
| **flaw** | Defekt *m*, Fehler *m* | **shortcoming** | Unzulänglichkeit *f* |
| **fluctuate** | schwanken *v* | **store** | lagern *v* |
| **interact** | aufeinanderwirken *v* | **track** | verfolgen *v* |
| **invoice** | Rechnung stellen *v* | **utilize** | verwenden *v* |
| **maintain** | instandhalten *v* | **variation** | Veränderung *f* |
| **noise** | Lärm *m* | **view** | Aussicht *f*, Blick *m* |

# Exercises XXIII

### 1. Answer the following questions:

a) How are the activities of a manufacturing enterprise controlled today ?

b) In what way does the modern approach to manufacturing differ from that of the past ?

c) What is a CAD system composed of ?

d) Where is the data generated in a CAD process stored ?

e) Why is it advantageous to combine CAD and CAM into a common process ?

f) What manufacturing tasks are controlled by MRP ?

g) What is meant by the term releasing ?

h) What is meant by computer integrated manufacturing (CIM) ?

i) What kinds of operations are carried out in a scheduling process ?

j) What is meant by the expression inventory on hand ?

### 2. Fill in the gaps in the following sentences:

a) The _____ approach to manufacturing was to view it as a sequence of _____.

b) Detailed _____ design was done by making drawings on a _____.

c) The modern design method which is called _____ uses a computer with _____.

d) The object which is being _____ can be viewed on a _____.

e) The proper use of computers _____ the manufacturing process, and increases the _____ of an enterprise.

f) Releasing is the _____ of initiating _____.

g) These requirements are _____ with the _____ on hand.

h) An increase in ____ can be achieved by combining ____.

i) Lead time is the time taken between ____ of an order and the ____ of the product.

j) Reporting is the process of ____ management about how ____ is proceeding.

**3. Translate into English:**

a) Wettbewerbsdruck hat die Unzulänglichkeiten dieser Ansicht offengelegt und Hersteller dazu gezwungen, einen systematischen Ansatz anzunehmen, in dem die verschiedenen Aktivitäten Teile eines dynamischen interagierenden Systems bilden.

b) Die Detailauslegung eines Produkts wurde in der Vergangenheit durch das Erstellen von Zeichnungen auf einem Zeichenbrett durchgeführt. Die moderne Verfahrensweise, die rechnerunterstützter Entwurf genannt wird, verwendet einen Computer mit geeigneter Software.

c) Durch den Einsatz von geometrischer Modellierung und Analyse können verschiedene Auslegungsvarianten untersucht werden. Ein besonderer Vorteil dieser Methode ist, dass das Objekt aus verschiedenen Winkeln am Bildschirm betrachtet werden kann.

d) Eine Steigerung der Effizienz kann durch die Integration von CAD und CAM in ein gemeinsames CAD/CAM Verfahren verwirklicht werden. Eine gemeinsame Datenbank ist notwendig, um die Automatisierung des gesamten Prozesses von der Auslegungsphase bis zur Herstellungsphase zu ermöglichen.

## 24 The control of manufacturing costs

The ultimate criterion of the value of a product is its *selling price*. This price includes not only the *direct* or *manufacturing cost*, but also many *indirect* or *overhead costs* like administration, sales, research and development, maintenance, etc. The manufacturing costs can be reduced and controlled by the right choice of materials, design, and manufacturing processes. Indirect costs however may be *more difficult to control*. Careful and constant monitoring of costs is required, if the *total cost* is to be kept within manageable limits.

**Unit costs and productivity**

The *unit cost* of a product can be calculated and fixed without much difficulty. *Productivity* is however a much more difficult quantity to define and measure. It is usually defined as the *value* of goods *produced* per *employee*. This definition refers to *labour productivity* and does not take into account the amount of capital employed. *Capital productivity* which is even more difficult to measure, has become important with the increase in automation, and the consequent decrease in the number of workers employed. *Automation* calls for heavy investment in machines, and is only profitable where large quantities of goods are produced. Unit costs are reduced by automation, but an *economic limit* is very often reached, after which further investment yields *diminishing reductions* in unit costs.

**Total cost**

The total cost may be expressed in terms of the individual costs as follows:

| Total cost = direct costs + indirect costs + fixed costs |
| --- |

**1. Direct or operating costs**

Direct costs refer to the *actual costs* of *manufacturing,* and are proportional to the number of units produced. Such costs can be divided into two parts, *net material costs* and *labour costs.* The term net material cost refers to the cost of the raw material *minus the cost* of the remaining *scrap material*. The raw material may be in many forms like sheet metal, powder material, or even in the form of castings or forgings obtained from an outside supplier.

*Labour costs* refer to the cost of the personnel employed. Both these costs can be calculated per unit produced. Additional production costs like the cost of *energy* used in the production processes, and the cost of *special tools* and *dies* should be included.

Clearly the contribution made to the unit price by tooling costs are reduced if a large number of units are produced. For example, the cost of a die casting mould may be 10000 dollars. If 20000 units are produced using this mould, the *added cost* due to the mould is half a dollar per unit. The division of costs into direct and indirect costs is a *fuzzy* one. What is important is that *all costs* be included in the *cost calculations.*

## 2. Indirect costs or overhead costs (also called burden)

Indirect or overhead costs may be greater than direct costs. Indirect costs are the costs of the *functions* and *services* which are *additionally required* for the efficient running of a manufacturing enterprise. Some of these are the following:

a)  Indirect labour services which include material handling and transfer, repair and maintenance, supervision, cleaning, etc.

b)  Engineering services like research and development, quality control, laboratory testing, etc.

c)  Energy costs like heating and lighting costs that are *not directly used* in production.

d)  Sales and marketing services.

e)  Administration and management services.

The manufacturing processes can be carefully analyzed, improved, and made more efficient. The cost savings due to any *improvement in productivity* may be *nullified* however, if indirect costs are not carefully controlled. This requires the *breakdown* of the *indirect costs* into their elements, and subjecting each element to a careful and rigorous analysis.

## 3. Fixed costs

Fixed costs include the costs of buildings, equipment, and all other facilities in general. In calculating fixed costs, depreciation, interest, taxes, and insurance must be taken into account. When manufacturing costs are estimated, fixed costs are calculated on the basis of *expected equipment utililization*. Thus the fixed cost per unit is halved if a machine is used for two shifts per day instead of one shift per day.The fixed costs are usually expressed in terms of a *machine-hour rate* or burden.

## Quantity of units produced

The unit cost will clearly be less when large quantities are produced. It will also be clear that a *minimum number* of *units* will have to be produced before a profit can be made. This minimum number of units is known as the *break-even number*. Fig 24.1 shows how the different costs vary with the number of units produced, and also the *profits* made *before* and *after taxes*.

## Outsourcing of supplies

Every manufacturer has to consider whether it is not better to buy a component or a product from another supplier rather than manufacturing it himself. Quite often there are *specialized suppliers* who can produce a component of higher quality cheaper than the manufacturer himself. Automobile manufacturers buy a large percentage of their components from *outside suppliers*. The manufacture of the entire range of components that they use would require more capital, more expertize, and more production facilities than their resources would allow.

# Vocabulary

| | | | |
|---|---|---|---|
| **administration** | Verwaltung *f* | **machine-hour rate** | Machinenstundensatz *m* |
| **break-even number** | Rentabilitätsgrenze *f* | **manageable** | beherrschbar *adj* |
| **costs** | Kosten *pl* | **material handling** | Fördertechnik *f* |
| **criterion** | Maßstab *m*, Kriterium *n* | **monitor** | überwachen *v* |
| **customer** | Kunde *m* | **nullify** | anullieren, vernichten *v* |
| **diminish** | vermindern, verringern *v* | **net** | netto *adj* |
| **element** | Grundbestandteil *m* | **refer** | verweisen, hinweisen *v* |
| **employ** | beschäftigen *v* | **remain** | übrigbleiben *v* |
| **expertize** | Sach-, Fachkenntnis *f* | **rigorous** | streng, sorgfältig *adj* |
| **facility** | Einrichtung *f*, Anlage *f* | **sale** | Verkauf *m* |
| **fixed costs** | Anlagekosten *pl* | **service** | Dienstleistung *f* |
| **fuzzy** | undeutlich *adj* | **tax** | Steuer *f* |
| **improve** | verbessern *v* | **ultimate** | endgültig *adj* |
| **include** | einschließen, enthalten *v* | **unit** | Einheit *f* |
| **limit** | Grenze *f* | **utilize** | verwenden *v* |
| **labour costs** | Arbeitskosten *pl* | **value** | Wert *m* |

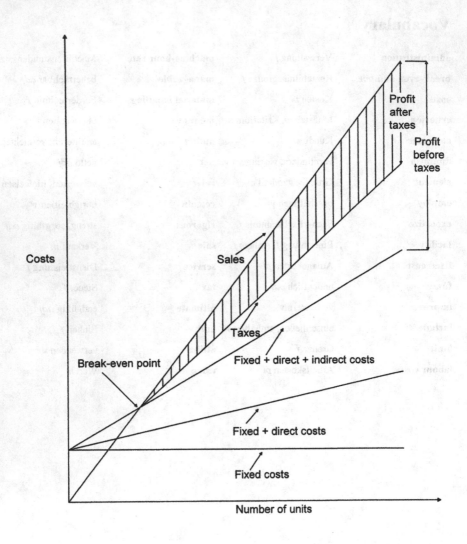

Fig 24.1 The relationship between sales, costs, and profits

# Exercises XXIV

## 1. Answer the following questions:

a) How is productivity usually defined ?

b) What is meant by direct costs ?

c) What does the term net material costs refer to ?

d) What kinds of energy costs are included in the indirect costs ?

e) Why must indirect costs be carefully controlled ?

f) What types of costs are included in the calculation of fixed costs ?

g) What is meant by the term break-even number ?

h) What costs are proportional to the number of units produced ?

i) What is meant by the term outsourcing ?

j) Why don't automobile manufacturers manufacture all the components that they use ?

## 2. Fill in the gaps in the following sentences:

a) The ultimate _____ of the value of a product is its _____ .

b) Productivity is usually _____ as the value of _____ .

c) The total cost can be divided into _____ , indirect costs, and _____ .

d) Direct costs are _____ to the number of _____ .

e) The division of costs into _____ and indirect costs is a _____ .

f) The improvement in productivity may be _____ if indirect costs are not carefully _____ .

g) The _____ per unit is halved if a machine is used for _____ a day.

h) Specialized _____ can produce a component _____ than the manufacturer himself.

i) Automobile manufacturers buy a _____ of their components from _____ .

j) What is important is that all costs be _____ in the cost _____.

## 3. Translate into English:

a) Die Stückkosten eines Produkts können ohne größere Schwierigkeiten geschätzt werden. Produktivität ist jedoch eine sehr viel schwerer zu messende Grösse. Sie wird üblicherweise definiert als Produktionsgegenwert pro Beschäftigtem.

b) Automatisierung verlangt eine große Investition in Maschinen, und ist nur profitabel, wenn große Mengen Güter hergestellt werden. Stückkosten werden durch Automatisierung gesenkt, aber es wird eine ökonomische Grenze erreicht, jenseits derer weitere Investitionen eine verschwindend kleine Verringerung der Stückkosten erzielen.

c) Direkte Kosten beziehen sich auf die tatsächlichen Herstellungskosten und sind proportional zu der Anzahl der hergestellten Einheiten. Solche Kosten können in zwei Teile, Materialkosten und Arbeitskosten, eingeteilt werden.

d) Das Rohmaterial kann in vielen Formen vorliegen wie Metallblech, körniges Material oder sogar in Form von Gussstücken oder Schmiedestücken, möglicherweise von einem Drittlieferanten bezogen. Arbeitskosten beziehen sich auf die Kosten des beschäftigten Personals. Beide Kostenarten können pro hergestellte Einheit berechnet werden.

e) Jeder Hersteller muss bedenken, ob es nicht besser ist, eine Komponente oder ein Produkt von einem Drittlieferanten zu kaufen, der die Komponente in höherer Qualität günstiger als der Hersteller fertigen kann.

## Answers to exercises

# Exercises I

**1.**

a) Wrought iron has the advantage that it can be easily bent and formed into different shapes. It has also the advantage that it does not rust easily.

b) Steel which contains about 0.2 % carbon is called mild steel. It is used for many purposes, such as for reinforcing concrete, for making car bodies, etc.

c) Carbon steel contains from 0.3 to 1.5 % carbon and can be hardened. However it loses its hardness easily at high temperatures, and therefore tools made from it are unsuitable for use in metal cutting machines.

d) Alloy steels contain additional alloying elements like tungsten, vanadium, and chromium, and can work at higher metal cutting temperatures than carbon steel.

e) Cast iron is composed of iron with about 3 % of carbon. When melted, it flows easily into moulds, enabling objects with complex forms to be cast from it.

f) The nonferrous metals which are most commonly used as engineering materials are aluminium and its alloys, copper and its alloys, and zinc alloys.

g) Goods made from plastics have the advantages of being cheap, light, and easy to manufacture.

h) Thermoplastics soften when heated and become hard on cooling. They can be moulded repeatedly. Thermosetting plastics are harder than thermoplastics, but can be heated and moulded only once.

i) A good example of a composite material is resin reinforced with fibreglass. It has the advantages of high strength and easy mouldability at room temperature.

j) The two types of silicon n and p are manufactured by adding small amounts of impurities to very pure silicon.

**2.**

a) Different kinds of metals and alloys are required for different purposes.

b) Wrought iron has the advantage of being easily bent into different shapes.

c) Steel is an alloy composed of iron and a small amount of carbon.

d) Carbon steel can be hardened, and is used to make small tools like scissors and screwdrivers.

e) Metal cutting tools are made of special alloy steels called high speed steels.

f) Tools made of alloy steels can work at higher temperatures than tools made of carbon steel.

g) Cast iron when melted flows freely into moulds.

h) Thermoplastics become soft when heated and can be moulded into different shapes.

i) Composite materials can be made from resins reinforced with glass fibre.

j) Mild steel can be used for reinforcing concrete and for making car bodies.

**3.**

a) Metal alloys and plastics are amongst the most important materials used by manufacturing industry today. Different applications require different metals and alloys.

b) Mild steel which contains about 0.2% carbon, is the most used type of steel, and is used for many purposes, like the reinforcing of concrete and the construction of water tanks. Mild steel, unlike steels with a high carbon content cannot be hardened.

c) Cast iron is a very useful material, because in liquid form it is able to flow easily into moulds. Complex shapes can also be cast using cast iron. Cast iron is a hardwearing material which resists abrasion.

d) Plastics are of increasing importance, because goods made of plastics are easy to manufacture, reasonable in price, and light. Some plastics become soft when heated, and hard when cooled. Such plastics can be used repeatedly.

e) Composite materials which are made of resin and glass fibre are used when high strength and easy moulding at room temperatures, are some of the properties required.

# Exercises II

**1.**

a) Strength, hardness, ductility, and toughness are four of the most important mechanical properties of a material.

b) Some of the most important mechanical properties of a material can be evaluated by carrying out a Young's modulus experiment.

c) Longitudinal stress = Tensile force / cross-sectional area, and longitudinal strain = Increase in length / original length.

d) The strain is proportional to the stress in the elastic region.

e) Necking is said to occur when the cross-section of a specimen decreases only at its weakest point.

f) In structural applications, loads should be kept well within the elastic range.

g) Ductility refers to the property of a metal which allows it to be plastically deformed and to be drawn into wire without breaking.

h) A brittle material breaks when plastically deformed.

i) Toughness refers to the ability of a material to withstand bending without breaking.

j) Hardness refers to a material's ability to resist indentation or abrasion.

**2.**

a) Foremost amongst the properties that should be considered are mechanical properties.

b) The load is increased from zero until the specimen breaks.

c) If the load is removed, the specimen will revert to its original state.

d) The stress at the maximum load is called the ultimate tensile strength.

e) Elastic and plastic deformation are both useful in engineering applications.

f) Toughness refers to the ability of a material to withstand bending.

g) Ductility refers to the property of a material which allows it to be drawn into wire.

h) Creep refers to a phenomenon which usually occurs at high temperatures.

i) Properties which facilitate manufacture need also to be considered.

j) Hardness refers to the ability of a material to resist indentation.

**3.**

a) An important way of assessing the main mechanical properties of a material is by carrying out an experiment to see how its length changes, when it is subjected to a tensional load.

b) At higher values of stress beyond A, the slope of the curve changes, and the deformation is no longer proportional and elastic. The specimen does not return to its original shape and size when the load is removed.

c) In structural applications like columns and machine frames, loads must be kept well within the elastic range, so that no permanent deformation occurs.

d) Ductility refers to the property of a metal which allows it to be drawn into wire without breaking. A ductile material must have both strength and plasticity. Lead for example is ductile, but is difficult to draw into wire because its strength is low.

e) Hardness refers to the ability of a material to resist abrasion and indentation. The hardness of a material can be measured by a hardness test such as a Brinell test.

# Exercises III

**1.**

a) Manufacturing industry produces goods today using machines, computers, and automation in such an efficient manner, that the need for hand tools has been almost eliminated.

b) Hand tools are still needed for the making of prototypes and models, and for the repair and maintenance of equipment.

c) The term maintenance refers to the work that needs to be done to keep machines and equipment in good working condition.

d) Chisels have to be ground to a definite shape, so that they work efficiently and do not become too hot.

e) Taps and dies are used for cutting internal and external screw threads.

f) An offset screw driver is used to reach screws in places which cannot be reached by an ordinary screw driver.

g) A ring spanner surrounds the head of a bolt completely, and therefore has a much firmer grip on the head of a bolt than an ordinary spanner which surrounds the head only partially.

h) A try square is used to check if two surfaces are perpendicular to each other.

i) Vee-blocks are used to hold cylindrical rods or pipes while work is being done on them.

j) Measuring devices are required during the marking-out process, and also for checking the dimensions of a work piece.

**2.**

a) A hammer can be used for shaping and forming sheet metal.

b) Pliers are used for many purposes, like gripping and holding.

c) A spanner is used for tightening nuts and bolts.

d) Files are used to remove metal from the surface of an object.

e) The important angles of a chisel are the rake angle and the clearance angle.

f) Taps are used for cutting internal screw threads.

g) Angle blocks have two surfaces which are perpendiculatr to each other.

h) Measuring devices are used to check the dimensions of objects.

i) Try squares are used to check if two surfaces are perpendicular to each other.

j) Inside and outside calipers are used by machine tool operators.

**3.**

a) Although most goods are manufactured today using automatic machines, hand tools are still required for repair and maintenance work on machines and installations.

b) A hammer is a multi-purpose tool. One of its uses among others, is to hammer a piece of metal until it acquires the required shape.

c) An offset screw driver is a special tool. It is used for reaching screws in awkward places.

d) There are many types of spanners. When a bolt has to be gripped firmly, one uses a ring spanner or a socket spanner.

e) A metal saw which is used for cutting sheet metal is called a hacksaw. A piece of metal can be made to acquire a desired shape by working on it first with a chisel, and then with a file.

f) Measuring instruments are used for the marking out process, and also for checking the dimensions after the work has been done.

g) A try square is used to check if two surfaces are perpendicular to each other.

# Exercises IV

**1.**

a) Metal fasteners are used to join metal sheets or metal components by mechanical means.

b) Rivets are usually used when metal sheets have to be joined together permanently.

c) Fasteners with screw threads are used to join components which have to be taken apart later.

d) An additional part like a spring washer or a lock nut may be used to prevent a bolt and nut from becoming loose.

e) Set screws are usually made with a hexagonal hole in their head, so that they can then be tightened efficiently by using a hexagonal key.

f) Studs have to be used to join parts to cast iron components.

g) If a bolt is tightened excessively into a cast iron screw thread, the thread can be damaged.

h) A good example is the use of studs to hold together the cylinder head and the cylinder block of an automobile engine.

i) Studs are used in applications where water-tight joints are required.

j) It is not possible to use bolts and nuts as fasteners when only one side of the parts to be joined is accessible.

**2.**

a) Metal components can be joined together by using metal fasteners.

b) Riveting is used on ships' hulls and aircraft fuselages.

c) Metal parts may also be joined by fastening devices which have a screw thread.

d) When both sides of the parts to be joined are accessible, nuts and bolts may be used.

e) It is often necessary to use set screws with countersunk heads.

f) Countersunk screws often have a hexagonal hole in their head.

g) Studs are used in applications where heavy pressures are encountered.

h) Studs are used to join components to cast iron because the tensile strength of cast iron is very low.

i) The screw thread in a cast iron component may crumble if tightened excessively.

j) It is necessary to remove the head of a motor car engine when the engine needs overhauling.

3.

a) Metal components and sheets can be joined in many ways. One way is by the use of metal fasteners like rivets or bolts and nuts.

b) Bolts and nuts can be used when both sides of the components to be joined are accessible. An additional part like a washer could become necessary if the parts are subject to vibration.

c) Bolts with normal heads can be used to join metal components. However, it is often necessary to use bolts with countersunk heads. Such bolts have a hole of hexagonal shape in the head which enables the bolt to be tightened very effectively using a hexagonal key.

d) Studs are used in applications where heavy pressures are encountered and also where water-tight or gas-tight joints are needed.

e) Studs are used to join the cylinder head of an automobile engine to its cylinder block. The joint between the cylinder and the head has to be a temporary one, because it is necessary to remove the head when the engine needs reconditioning.

# Exercises V

**1.**

a) Soldering is the process of joining two metal objects by a third soft metal alloy called solder.

b) Solder is an alloy which melts at a lower temperature than the metals being soldered. Soft solder and hard solder are the commonly used types.

c) The surfaces of the metals must be thoroughly cleaned before soldering.

d) The types of flux usually used in a workshop are acidic in character.

e) Acid fluxes are too corrosive and resin is usually used instead.

f) In the brazing process, much higher temperatures are reached and the solder used is a hard solder called spelt.

g) In oxy-acetylene welding a blowpipe through which acetylene and oxygen gases flow in suitable amounts is used.

h) In the electric arc welding process, the metal electrode itself acts as a filler rod making the process easier to control.

i) In the resistance welding process, no filler rods or electrodes are required. Small portions of the metals being welded melt.

j) A large amount of heat is generated at the contact faces because the electrical resistance to the flow of current is large.

**2.**

a) The process of joining two metal surfaces by a third soft metal alloy is called soldering.

b) Hard solder is an alloy of copper and zinc and is called spelt or silver solder.

c) Flux removes oxides and grease from the metal surface and allows the solder to flow freely.

   d) Fluxes used in the workshop are usually acidic in character, and are corrosive.

   e) The solder used for electrical work is in the form of a wire with a resin core.

   f) The hard soldering process uses spelt as the solder and borax as the flux.

   g) The two principal types of welding are electric arc welding and resistance welding.

   h) In oxy-acetylene welding a blowpipe through which oxygen and acetylene gases flow through is used.

   i) The filler rod melts and a small amount of liquid metal is formed at the joint.

   j) In the electric arc welding process, the metal electrode also acts as a filler rod.

**3.**

   a) Solder is a metal alloy which melts at a somewhat lower temperature than the metals to be joined. Two types of solder are normally available. One type of solder which is an alloy of tin and lead is softer than the other which is an alloy of copper and zinc and is called silver solder.

   b) It is very important that the surfaces of the metals that have to be soldered should at first be thoroughly cleaned. For this purpose a flux which removes oxides and grease from the surface is used.

   c) In the hard soldering process, another solder and more heat must be used to reach the melting point of hard solder which is about 600°C. A hard soldered joint is considerably stronger than a soft soldered one.

   d) The temperature of the flame in an oxy-acetylene blow pipe can reach a maximum value of about 3000°C. The blowpipe heats the metal parts that have to be joined and also a filler rod which is held in the flame at the joint.

# Exercises VI

**1.**

a) Casting has the advantage of being probably the quickest and the most economical way of producing a metal component.

b) In sand casting, the molten metal is poured into a hollow space in a box filled with sand. The hollow space has the shape of the object to be cast.

c) A pattern is a replica of the object, and is usually made of wood or steel.

d) The four main stages in the manufacture of a casting are the making of the pattern, the making of the mould, the pouring in of the metal, and the removal of the casting from the mould.

e) The casting is usually made slightly larger than the final component that has to be produced so that it can be machined to precise dimensions.

f) Sand casting is particularly useful for producing large complex castings, like machine beds, and engine cylinder blocks.

g) The two main die casting processes are gravity die casting and pressure die casting. Die castings are stronger and better finished than sand castings made from the same metal or alloy.

h) Steel is used for the construction of die casting moulds.

i) The die casting process has the disadvantage that it can only be used with low melting point metals and alloys.

j) Many automobile components like carburettors, fuel pumps, and engine cylinder heads, are manufactured by the die casting process.

**2.**

a) The process of sand casting has been practiced for thousands of years.

b) In the sand casting process the molten metal is poured into a mould.

c) The pattern which is a replica of the object to be produced is made of wood or metal.

d) The wooden pattern is usually made in two halves which are held together by dowel pins.

e) The pattern and the resulting casting are usually made larger than the final size of the finished object.

f) The extra allowance in the size of the casting, is called a machining allowance.

g) The two methods of producing die castings are called gravity die casting and pressure die casting.

h) Die casting moulds are very precisely made and the castings produced are very accurate in their dimensions.

i) The disadvantage of the die casting process is that only low melting point metals and alloys can be used.

j) In the pressure die casting process the molten metal is forced into the dies under pressure.

**3.**

a) The process of casting is probably the quickest and easiest way to produce metal objects, particularly large and complex objects.

b) In the sand casting process, the melted metal is poured into the hollow space of a container filled with sand. The hollow space is similar in shape and size to the object that has to be cast.

c) The pattern and the casting are usually somewhat larger than the finished metal component. This extra allowance is called a machining allowance.

d) Moulds which are used in the pressure die casting process are made of steel and can be used indefinitely. The moulds are precisely made and the castings produced are very accurate in their dimensions.

e) The disadvantages of the pressure die casting method is that only low melting point metals and alloys, like aluminium and zinc alloys can be used in this process. However castings produced by this process have the advantage that no machining is necessary.

## Exercises VII

**1.**

a) Steel is unique in its ability to exist both as a soft material which can be easily machined, and after hardening as a hard material out of which metal cutting tools can be made.

b) Polycrystalline materials consist of a large number of very small crystals interlocked together.

c) Castings that are allowed to cool slowly have a coarse grain structure, while those that are allowed to cool rapidly have a fine grain structure.

d) Plastic deformation causes the grains in a metal to become elongated.

e) Stress relieving uses less energy than other processes, and has the advantage that the metal surface is not spoilt by oxidation or scaling.

f) Annealing softens steel, removes stress, and improves the toughness, ductility, and machinability of the metal.

g) Metal cutting tools and components that have to resist wear need to be hardened.

h) Hardened steel components must be tempered, because otherwise they would be too brittle for normal use.

i) Case hardened components have a coarse grain structure. They have to be refined to improve the grain structure and toughness of the core.

j) The surface of the component becomes hard due to the formation of hard nitrides on the surface.

**2.**

a) Slow cooling as in a sand casting results in a coarse grain structure.

b) Metals which undergo plastic deformation become work hardened.

c) The stress relieving process is relatively cheap, and has the advantage that it does not cause the metal surface to become oxidized.

d) The annealing process softens steel and increases its toughness.

e) Hardened steel is used to make components which resist wear.

f) Steel is hardened by heating it to a temperature above the upper critical line and then quenching in water.

g) Tempering reduces the brittleness and hardness of steel while increasing its toughness.

h) Case hardening produces a component with a hard surface and a tough interior.

i) Hardened steel acquires a needle like grain structure called martensite which makes it glass hard.

j) Carbon is absorbed into the surface converting it into a high carbon steel.

**3.**

a) The heat treatment process improves the mechanical properties of a metal to the extent that the manufacturing process is made easier. In addition, a heat treated metal component is able to withstand rough usage.

b) The mechanical properties of a metal like hardness, strength, and brittleness, depend on the size, shape, and orientation of the crystal grains in it.

c) When a casting is produced, the grain structure of the casting depends on the rate at which it cools when it is solidifying. If the rate of cooling is slow, as in a sand casting, the crystal grains are large. On the other hand, if the rate of cooling is rapid, the crystal grains are somewhat smaller.

d) Metals become highly stressed during the manufacturing process. This stress must be removed, before the component is put into use. This can be accomplished by heating the component to a temperature of above 500°C, and allowing it to cool.

# Exercises VIII

**1.**

a) Casting is probably the cheapest way of producing metal components.

b) Forged components are stronger and less brittle than cast components.

c) No material is wasted in the forging process because the entire piece of material is forced into the shape of the final component.

d) Cast iron cannot be forged because it becomes very brittle and breaks easily when heated to red heat.

e) A blacksmith produces forged components by using hand tools.

f) Mechanical hammers were used for forging after the beginning of the industrial revolution. They have the advantage of being able to forge large components.

g) Three ways in which a forging can be produced are hand forging, drop forging, and upset forging.

h) The drop forging process is particularly useful when a large number of medium-size forgings are required.

i) In the drop forging process, the red hot metal billet is forced into the cavity between the dies.

j) In the upset forging process, the cross-sectional area of the component is changed.

**2.**

a) Metal components can be made cheaply and efficiently by the casting process.

b) Cast components are usually brittle and break easily.

c) Forged components are stronger than cast components or those produced by machining.

d) Forging is more economical in the use of material, because the metal is forced into the required shape.

e) Mild steel becomes ductile at high temperatures, and can be forced into the right shape.

f) Forged components were produced in small quantities by blacksmiths.

g) Large mechanical hammers were used to forge large components.

h) Two half dies are used in the drop forging process, and the red hot billet is forced into them.

i) Often, several stages of forging are needed to produce a forged component.

j) In upset forging, the cross-sectional area of the component is changed.

**3.**

a) Metal components can be produced more cheaply by casting than by forging. Components produced by casting however are very brittle, and break easily.

b) The process of forging is very economical in the use of material, because the metal is pressed into the form of the final component, and no metal is wasted.

c) Steel and many metals become ductile at high temperatures, and can be pressed or hammered into the desired form. Cast iron cannot however, be forged in this way.

d) The method of drop forging is particularly useful, when a large number of components need to be produced.

e) In this process two half dies are used. The pressure exerted by the upper die, forces the red hot billet into the cavity between the dies.

# Exercises IX

**1.**

a) Metals are initially produced by the refining of ores or scrap in furnaces.

b) This transformation is usually necessary because intermediate products are the starting point for the manufacture of more complex products.

c) This is the hardening that takes place when a metal is plastically deformed.

d) The metal can be brought back into a normal state by heat treatment.

e) Smaller forces are required to cause deformation in a hot working process because the metal flows more easily at higher temperatures.

f) The thread rolling process produces stronger screw threads.

g) Thin metal sheets are produced from hot rolled sheets by cold rolling.

h) Screws, nails, and bolts are some of the products that can be manufactured from metal wire.

i) Hot rolled plates have rough surfaces and poor dimensional tolerances.

j) Cold working processes are able to manufacture products that have a better finish, closer tolerances, and thinner walls than those manufactured by the use of hot working processes.

**2.**

a) Metals are initially produced by the refining of ores and scrap metal in furnaces.

b) Stress is required to plastically deform a metal, and this causes strain or work hardening.

c) The disadvantages of hot working are that dimensional tolerances are low, and that the surface is spoilt by oxidation.

d) Ductility is high, and large deformations without fracture are possible.

e) The increase in strength due to strain hardening may be retained if required.

f) In flat rolling, the thickness of a slab is reduced to produce a longer product.

g) Hot rolled sheets have rough surfaces, and poor dimensional tolerances.

h) In ring rolling, pierced billets and centering mandrels are used to produce hollow products.

i) Smaller tubes below a certain diameter have to be cold drawn.

j) Rolling is the most important of the bulk deformation processes.

**3.**

a) Before metal goods can be manufactured, it is necessary to transform these cast metal forms into intermediate products like metal sheets, wire, and rods, which are the starting point for the manufacture of more complex products.

b) Unlike sheet metal working where the changes in thickness are small, bulk deformation processes cause large changes in thickness, diameter or other dimension of the cast metal forms.

c) In the drawing process, the material is pulled through a die of gradually decreasing cross-section. Wire is the starting point for the manufacture of a number of products like screws, nails, bolts, and wire frame structures.

d) Long rods, tubes, etc. of uniform cross-section can be produced by extrusion. In this process, the material is under pressure, and is forced to flow through a die. The cross-sections of the extruded products have different shapes and sizes, depending on the shape and size of the opening in the die.

# Exercises X

**1.**

a) In rotatory motion, the object is rotating round an axis or a point.

b) In reciprocating motion, the object moves along a straight line in one direction, and then back along the same straight line to its starting point.

c) A shaft is usually a cylindrical rod which undergoes a rotatory motion.

d) Bearings are needed as supports in which a shaft can rotate.

e) Bronze or white metal are suitable materials for making plain bearings.

f) A split bearing is a plain bearing which is made in two halves.

g) The bearing on the small end of a connecting rod is a single piece hollow cylindrical bearing.

h) A ball bearing consists of hardened steel spheres running between two precision ground hard cylindrical races.

i) The purpose of lubrication is to maintain a thin film of oil between two contact surfaces.

j) Two rotating shafts can be coupled together by using gear wheels, by using a belt and pulleys, or by using a chain and sprocket wheels.

**2.**

a) A rotating body is usually rotating round a point or an axis.

b) A body has linear motion when it moves in a straight line in one direction.

c) The supports in which a shaft rotates are called bearings.

d) The big end bearing of an automobile connecting rod consists of two thin split bearings.

e) The spheres in a ball bearing are fitted between two cylindrical races.

f) The process of maintaining a film of oil between two contact surfaces is called lubrication.

g) Two shafts may be coupled using a belt and two pulleys.

h) The small end bearing on an automobile connecting rod is called a gudgeon pin.

i) The rollers in a roller bearing are located between an outer cup and an inner cone.

j) Two shafts may be coupled using a chain and two sprocket wheels.

**3.**

a) When a body has a rotatory motion, it usually rotates round an axis in a clockwise or an anticlockwise sense. If the motion is linear the body moves in a straight line.

b) A machine is composed of many parts, some of which are moving. A shaft is a cylindrical rod which is usually made of steel and is supported by bearings in which it can rotate.

c) Bearings are usually in the form of plain bearings, ball bearings, or roller bearings. Special alloys are used to make plain bearings, for example bronze.

d) Ball bearings are composed of steel balls located between two cylindrical races. The lubrication of such bearings is simple because they only need an occasional application of grease.

e) A rotating shaft has mechanical energy. It is often necessary to transfer this energy to another shaft. This can be done by using two gear wheels or by using two pulleys and a belt.

# Exercises XI

**1.**

    a) Gears wheels are usually used to couple two or more rotating shafts.

    b) The ratio of the speeds between two shafts can be changed, by changing the ratio of the number of teeth in the gear wheels. The direction of the axes of rotation can be changed by using special gears like spiral gears.

    c) Spur gears have teeth which are cut parallel to their axes of rotation. They are used to couple parallel shafts.

    d) Bevel gears have teeth which lie on a conical surface and appear to meet at the apex of a cone.

    e) Helical gears have teeth which lie on the surface of a cylinder and are inclined at an angle to the axis of rotation of the gears.

    f) Spur gears are simple and cheap to make, but can only transmit motion between two parallel shafts. Bevel gears are able to transmit motion between two shafts inclined at an angle to each other, but are more expensive to make.

    g) Helical gears are placed in an oil bath to minimize wear on their teeth.

    h) A rack and pinion can be used to convert rotatory motion into linear motion.

    i) Worm gears are used for heavy duty work where a large ratio of speeds is required.

    j) Gearboxes are used when several changes in the ratio of the speeds between two or more shafts are required.

**2.**

    a) Gear wheels are used to couple two or more rotating shafts.

    b) A spur gear has teeth cut parallel to its axis of rotation.

c) The larger gear wheel is called the gear, and the smaller gear wheel the pinion.

d) The teeth in a bevel gear lie on a conical surface and appear to meet at the apex of a cone.

e) The teeth in helical gears lie on a cylinder, and are cut at an angle to the axis of rotation of the cylinder.

f) Helical gears are usually placed in an oil bath in order to minimize the wear on the teeth.

g) A rack and pinion converts rotatory motion into linear motion.

h) Worm gears are used for heavy duty work where a large ratio of speeds is required.

i) Helical gears are quieter and smoother in operation than spur gears.

j) When two gears are coupled together, the ratio of the speeds of rotation of the shafts depends on the ratio of the number of teeth in the gears.

**3.**

a) A gear wheel when coupled with other gear wheels transmits motion from one part of a mechanism to another.

b) In helical gears, the friction and the resulting heat and wear generated, are greater than with other drives. These are often placed in an oil bath in order to reduce wear.

c) In a bevel gear, the teeth are cut in such a way that they lie on a conical surface and appear to meet at the apex of a cone.

d) The teeth of a helical gear are cut on a helical cylinder at an angle to the axis of the drive. Helical gears are quieter, and run more smoothly than spur gears.

e) When two gears which are attached to two shafts are coupled together, the ratio of the speeds of the shafts depends on the ratio of the number of the teeth on both gear wheels.

# Exercises XII

**1.**

   a) The pitch is the distance from a point on a screw thread to a corresponding point on the next thread, measured parallel to the axis of the screw.

   b) The lead is the distance advanced by the screw when it is rotated by one revolution.

   c) The major diameter is the largest diameter of the screw thread, and the minor diameter is the smallest diameter of the screw thread.

   d) The depth of engagement of a screw thread is the distance measured radially over which a mating male and female thread overlap.

   e) When a screw is used without a nut (as in a wood screw) it is called a screw, and when it is used with a nut it is called a bolt.

   f) Locking devices are used to prevent nuts from becoming loose.

   g) A key is a piece of steel inserted between a shaft and a hub, usually in an axial direction, to prevent rotation between them.

   h) The key is fitted into recesses called keyways which are cut in both the shaft and the hub.

   i) Spline shafts and hubs are used for heavy duty couplings. Such couplings allow an axial movement between the shaft and the hub.

   j) A cotter is a flat wedge shaped piece of steel which is used to fasten rods which are subjected to axial forces only.

**2.**

   a) Screws are used for fastening various types of components.

   b) The lead is the distance advanced by the screw when rotated through one revolution.

c) In a double thread screw, the lead is twice the pitch.

d) When a screw is used without a nut it is called a screw, and when it is used with a nut it is called a bolt.

e) A key is inserted between a hub and a shaft usually in an axial direction.

f) Recesses called keyways are cut into the shaft and hub to accomodate the key.

g) A feather key is a rectangular key which is fixed to the hub or the shaft.

h) The spline shaft has a number of key like projections spaced equally round its circumference.

i) A cotter is used to fasten rods which are subjected axial forces only.

j) It is necessary to use locking devices to prevent nuts from becoming loose.

**3.**

a) An external screw thread is cut on the outside of a cylinder, while an internal screw thread is cut on the inside of a bored hole, like in a nut.

b) There are many types of screws threads, each suitable for a particular type of work. Each screw thread is made to a definite specification and has a special name, for example a metric screw thread.

c) In a single thread screw, the pitch and the lead are the same. In a double thread screw, the lead is double the pitch.

d) Screws differ from each other in the dimensions of their screw threads, in the shapes of their heads, and in other details. Hexagonal screws are the screws that are most used in mechanical construction. Set or grub screws are headless screws which have a screw thread along the entire length of the shaft.

# Exercises XIII

**1.**

a) Engineering inspection is important, because it ensures that the components which are manufactured have the right dimensions.

b) Variations in dimensions must be kept within strict limits to ensure the interchangeability of parts.

c) Nominal dimensions are the dimensions given on the drawing of the component.

d) The deviation from a nominal dimension is called a tolerance. This is necessary, because it is impossible to produce a large number of components having precisely the dimensions given in the drawing.

e) The two limits between which the dimensions of a component lie are called the low limit and the high limit.

f) In a clearance fit, a shaft must be able to move freely in a hole without being loose in it.

g) When the allowed deviation or tolerance is only above or only below the nominal dimension, it is called a unilateral tolerance. If the deviation can be on both sides of the nominal dimension, then it is called a bilateral tolerance.

h) The engineering allowance is the difference in dimensions between the high limit of the shaft and the low limit of the hole.

i) Allowances and tolerances must be kept as large as possible, otherwise the number of rejected components will be high, and this will increase the manufacturing cost.

j) An interference fit occurs when a shaft is slightly larger than a hole, and can only be forced into the hole under pressure.

**2.**

a) Engineering inspection is an essential part of all manufacturing. However, it is not usually possible to inspect every component for dimensional accuracy.

b) Small deviations in dimensions must be kept within strict limits.

c) Limits are necessary to ensure that components are interchangeable.

d) The dimensions of a component given in the drawing are called nominal dimensions.

e) It is impossible to produce components which have precisely the dimensions given in the drawing.

f) A certain amount of deviation from the nominal dimensions must be allowed and this deviation is called a tolerance.

g) In a clearance fit, the shaft must be able to move freely in a hole without being too loose in it.

h) The engineering allowance is the difference between the high limit of the shaft, and the low limit of the hole.

i) Small tolerances increase the cost of manufacture, and also the number of rejected components.

j) If the shaft is made slightly larger than the hole, pressure is required to force the shaft into the hole.

**3.**

a) Although engineering inspection is an essential part of engineering production, it is not possible to inspect every manufactured component.

b) When a component is designed, it is given certain dimensions on the drawing which are called nominal dimensions. However, it is not possible to manufacture a large number of components that have precisely the dimensions given on the drawing.

c) For this reason, a definite deviation from a nominal dimension is allowed, and this is called a tolerance. When a component is manufactured, each dimension must lie between two values called the high and low limits.

d) For a shaft to be able to enter a hole, there must be a small difference between the upper limit of the shaft, and the lower limit of the hole. This difference is called an allowance.

e) In general, tolerances must be kept as large as possible. Small tolerances increase the manufacturing costs and also the number of rejected components.

# Exercises XIV

1.

   a) The low cost of producing high quality metal pressings makes press work one of the most attractive and important of manufacturing processes.

   b) The first step in the manufacture of sheet metal products is called shearing.

   c) Piercing refers to the operation of producing a hole of any shape, while punching refers to the operation of producing a circular hole.

   d) Trimming refers to the removal of excess material from a pressed object.

   e) In a perforation process, a regularly spaced pattern of holes is produced.

   f) In a notching operation, a piece of metal from the edge of a metal sheet is removed.

   g) The forward movement of the metal strip is controlled by a stop stud.

   h) A die set consists of punches and dies fitted with suitable guides and collars.

   i) The stripper plate prevents the metal strip from rising too far upwards.

   j) The concentricity of the two holes is ensured by the use of a pilot.

2.

   a) The first step in the manufacture of sheet metal products is called shearing.

   b) Trimming refers to the removal of excess material from a pressed object.

   c) The low cost of producing high quality metal pressings makes press work an attractive manufacturing process

d) Piercing is the operation of producing a hole of any shape in a sheet of metal.

e) Notching is the operation of removing a piece of metal from the edge of a metal sheet.

f) Perforating is the operation of producing a regularly spaced pattern of holes in a sheet of metal.

g) The blanking punch carries a pilot which fits into the already pierced hole.

h) Blanking can be done by using a press fitted with an appropriate die set.

i) The punches are slightly smaller than the holes in the die.

j) The metal strip is prevented from rising by the stripper plate.

**3.**

a) Metal working presses are used by industry for the production of a large number of articles from sheet metal. The low cost and the high quality of the metal pressings that can be produced by industry today, makes this process a very attractive one.

b) The blanking of articles from metal sheet can be done by using a metal press fitted with appropriate die sets. The die set shown in the figure consists of two punches and a die. The metal strip is fed in from the right up to the small stop stud. The punches move downwards through the stripper plate, and punch out two blanks from the strip.

c) Piercing is the operation of producing a hole of any shape in a sheet of metal using a punch and a die. The material removed is unimportant and is treated as scrap in contrast to blanking where the removed piece of metal is the desired article.

d) Blanking is the operation of cutting out a piece of metal of the desired shape by using a punch and a die. Perforating is the operation of producing a regularly spaced pattern of holes in a sheet of metal.

# Exercises XV

**1.**

a) A metal undergoes plastic deformation in a bending process.

b) Only ductile metals can be bent easily without breaking.

c) Special presses called press brakes are required to bend long lengths of metal.

d) The metal sheet is clamped round its circumference in the stretch forming process.

e) The thickness is very little changed in the deep drawing process.

f) The pressure pad has the function of ironing out any wrinkles.

g) Combination dies are used to carry out blanking and drawing simultaneously.

h) Plunging is an operation in which a punch is pressed through a hole in a metal sheet, bending it into the shape required to take the head of a screw.

i) Embossing is an operation in which figures, letters, or designs are formed on sheet metal parts.

j) The metal is stretched beyond the elastic limit but not so far that it cracks or fractures.

**2.**

a) The bending and forming of metals are cold working operations.

b) The metal has to be stretched beyond the elastic limit.

c) Only some metals have sufficient ductility to be used in this way.

d) Complex profiles can also be formed by passing the metal strip through successive sets of rollers.

e) In pure stretch forming the metal sheet is completely clamped round its circumference.

f) The blank is held firmly on the die by the pressure plate.

g) The punch moves downwards and pushes the blank into the cavity.

h) In the spinning process, a thin sheet of metal is formed into the desired shape by revolving it at high speed and pressing it against a former.

i) Short lengths of metals can be bent by using die sets in mechanical presses.

j) Longer lengths require special presses with long beds called press brakes.

**3.**

a) The metal has to be stretched beyond the elastic limit. Only some metals and alloys have sufficient ductility to be used in this way.

b) Short lengths of metal can be bent using dies in mechanical presses. Longer lengths of metal require special presses with very long beds. Complex shapes can be formed by repeated bending.

c) In stretch forming, the metal is completely clamped round its circumference. Stretching the metal causes it to become thinner. The change in shape is achieved at the expense of sheet thickness.

d) In the spinning process, a thin sheet of metal is revolved at high speed while a special tool presses it against a former attached to the headstock spindle of the lathe. The metal is also supported at the tailstock. The pressure applied by the tool forces the metal to acquire the shape of the former.

# Exercises XVI

## 1.

a) The accuracy of the work done on a lathe depends on the skill and experience of the operator.

b) A lathe is unsuitable for production work because a lot of time is taken for tool setting, tool changing, etc.

c) A lathe has a rigid bed with parallel guideways on which are mounted a fixed headstock and a movable tailstock. In addition there is a carriage which can be moved along the guideways of the bed.

d) The main component of the headstock is a strong spindle which is driven by a motor through a gear box.

e) The workpiece is usually held in a chuck.

f) The tool post holds the cutting tools and is fixed to the carriage.

g) Capstan and turret lathes have the same types of headstock and tool post as ordinary lathes. The tailstock is however replaced by a hexagonal turret.

h) The important angles in a lathe tool are the top rake angle, the side rake angle, the front clearance angle, and the side clearance angle.

i) A cutting fluid is used to remove the heat generated during the cutting process and consequently increase the cutting life of the tool.

j) Water soluble oils are cheap and efficient coolants and are suitable for the machining of most steel components. However, they do not have very good lubricating properties.

## 2.

a) Lathes can produce components that are symmetrical about an axis.

b) The accuracy of the work done on a lathe depends on the skill and experience of the operator.

c) The ordinary lathe is unsuitable for production work because a lot of time is spent on tool changing, tool setting, etc.

d) The lathe has a bed on which are mounted a fixed headstock and a movable tailstock.

e) The work to be machined is held in a chuck.

f) The work is turned down to the desired dimensions.

g) The headstock contains a strong spindle driven by an electric motor through a gearbox.

h) Many machine tools like lathes and shaping machines use single point metal cutting tools.

i) The cutting fluid helps to remove chips from the edge of the tool, and improves the surface finish.

j) It is absolutely important that the correct cutting angles be ground on the tip of the tool.

3.

a) A lathe can be used to turn cylindrical surfaces both external and internal, and also to turn conical surfaces. An additional feature of the lathe, is its ability to cut screw threads on a cylindrical surface.

b) The accuracy of the work done on a lathe depends on the skill and experience of the operator. A lot of time is required for tool setting and tool changing, with the result that this type of work is not suitable for production.

c) The lathe has a rigid bed with parallel guideways, on which are mounted a fixed headstock, and a movable tailstock. In addition, it has a carriage which can be moved along the guideways, in a direction parallel to the axis of rotation of the spindle.

d) The use of a cutting fluid removes the heat generated during the cutting process, and thus increases the life of the tool. The cutting fluid also helps to remove chips from the cutting edge of the tool.

# Exercises XVII

1.

a) A compound table is a table with two independent movements at right angles to each other. The work is moved using the lead screws on the table, until the hole is precisely located under the drilling head, after which drilling is done.

b) The different parts of a twist drill are the body, the shank, and the tang.

c) Small drills are held in a self-centering chuck.

d) A reamer is used to finish a hole accurately to size.

e) Counterboring opens the end of a hole cylindrically, while counter-sinking opens the end of a hole conically.

f) The tools used in a lathe are single point cutting tools, while the tools used in a milling machine are multipoint cutting tools.

g) A vertical milling machine has a head with a spindle which is perpendicular to the work table. The head can be swivelled at an angle.

h) Ordinary milling machines lack the rigidity required for heavy work.

i) Accessories like dividing heads, vertical milling attachments, rotary tables, etc. can be used with a universal milling machine.

j) These accessories enable the machine to produce gears, twist drills, milling cutters, etc.and do a variety of milling and drilling operations.

2.

a) A compound table has two independent movements at right angles to each other.

b) Small drills have straight shanks and are held in a self-centering chuck.

c) Larger drills have taper shanks and are fitted directly into the spindle of the drilling machine.

d) A hole can be enlarged to the correct size by using a reamer.

e) Counterboring opens the end of a hole cylindrically.

f) Milling machines use multipoint cutting tools and not single point cutting tools.

g) Cutting tools and arbors can be fitted into a hole in the spindle.

h) The spindle receives power from the motor through belts, gears, and clutches.

i) The use of multipoint cutting tools enables the milling machine to achieve fast rates of metal removal.

j) Milling machines are fitted with accessories like dividing heads and rotary tables.

**3.**

a) A drill is usually made from a piece of HSS steel and has spiral grooves. At the beginning of the drilling process, a centre drill is used to drill a small hole. Then a twist drill is used to drill the hole to the right size.

b) Small drills usually have cylindrical shanks and are held in a self-centering chuck. Larger drills have tapered shanks which enable them to be inserted into the machine.

c) These machines are extremely versatile and have three independent movements longitudinal, transverse, and vertical. They are used in workshops, but lack the stability required for heavy production work.

d) The spindle has a tapered hole into which various cutting tools and arbors can be inserted. The arbor extends the machine spindle, and milling cutters can be mounted on it.

e) In a vertical milling machine, the spindle head is perpendicular to the work table. The spindle head can also be inclined at an angle, thus permitting the milling of angular surfaces.

## Exercises XVIII

**1.**

a) In the grinding process, metal is removed from a metal surface by a rotating abrasive wheel.

b) Grinding produces high dimensional accuracy and a good surface finish.

c) Very little metal is removed from a surface by grinding.

d) Grinding is the only method that can be used to machine materials that are too hard to be machined by other methods.

e) The different grinding methods available are external and internal cylindrical grinding, surface grinding, and form grinding.

f) Surface grinding is carried out by using either the periphery or the end face of the grinding wheel.

g) The surface becomes strengthened by work hardening and also becomes fatigue resistant.

h) Buffing produces a mirror like finish unobtainable by polishing.

i) The barrel finishing process eliminates hand finishing, and is therefore very economical in the use of labour.

j) In the shot blasting process, particles of abrasives or other materials moving at high velocity are made to strike the surface being treated.

**2.**

a) Grinding is a finishing operation which gives high dimensional accuracy to workpieces.

b) Very little metal is removed from the surface in a grinding operation.

c) Surfaces which are too hard to be machined by other methods can be machined by grinding.

d) External cylindrical grinding is used to produce cylindrical or tapered external surfaces.

e) Internal cylindrical grinding is used to produce cylindrical holes or internal tapers on a workpiece.

f) In surface grinding, the periphery or the end face of the grinding wheel may be used.

g) Buffing is a refined kind of polishing in which a mirror finish is produced.

h) Shot peening is a process used to strengthen and harden a surface.

i) Superfinishing is a process which uses bonded abrasive stones to produce a surface finish of extremely high quality.

j) Barrel finishing eliminates hand finishing and is therefore very economical in the use of labour.

**3.**

a) Grinding is an operation which is done using a rotating grinding wheel, and which removes metal from the surface of a workpiece.

b) Grinding is normally a finishing process, which gives a good surface finish and high dimensional accuracy to workpieces which have been machined by other methods.

c) Materials which are too hard to be machined by other methods can be machined by grinding. Only a small amount of metal is removed in a grinding operation.

d) External cylindrical grinding is used to produce a cylindrical or tapered surface on the outside of a workpiece. The surface grinding process is used to generate flat surfaces, using the end face or the periphery of the grinding wheel.

e) The abrasives used in grinding wheels, are small grains of silicon carbide or aluminium oxide. Grinding wheels are made by using a suitable material to bond the abrasive particles together.

# Exercises XIX

**1.**

a) The two most commonly used types of plastic material are thermoplastics and thermosetting plastics.

b) These processes are cost-effective because products having complicated forms can be moulded in a single operation.

c) This is because of the high cost of the moulds used.

d) The injection moulding process is the most widely used process.

e) Semifabricated plastic products are produced by extrusion.

f) A screw feed mechanism is used to feed raw plastic granules into the plasticizing cylinder.

g) In the pressure assisted vacuum forming process, pressure is applied from above in addition to the vacuum applied from below.

h) Goods made from thermosetting plastics can be heated and moulded only once, while those from thermoplastics can be heated and moulded repeatedly.

i) Thermoforming processes are used to make open container like objects.

j) The product is cooled in a stream of air to ensure rapid hardening.

**2.**

a) The raw materials used in the manufacture of plastic goods are in the form of powder, granules, or liquid.

b) Plastic manufacturing processes are cost-effective because complicated shapes can be moulded in a single operation.

c) Granules of the raw plastic material are fed into a heated plasticizing cylinder through a screw feed mechanism.

d) The screw feed mechanism is pushed to force the soft plastic through an injection nozzle into a two-piece mould.

e) The plastic is heated, compressed, and degassed, until it is in a soft state.

f) In vacuum forming the sheet is clamped and a vacuum is applied from below.

g) The mould is cooled rapidly, and the plastic object hardens quickly.

h) In the blow moulding process, pieces of extruded soft plastic are pinched off and welded to the bottom of a die.

i) The moulded object is removed after cooling by separating the two halves of the die.

j) Objects made from thermosetting plastics are permanently hardened.

**3.**

a) Plastic manufacturing processes are very cost-effective because complicated shapes can be in moulded in a single process. However the moulds used are very expensive, and such processes are only profitable if large quantities are produced.

b) This is probably the most widely used process for the production of moulded plastic products. Granules of the raw plastic material are fed into a heated plasticizing cylinder through a screw feed mechanism.

c) One of the commonest ways of producing bar, tube, film, etc. from thermoplastic materials is by extrusion. The screw feed mechanism presses the soft plastic through an opening in a die which has the desired cross-section.

d) In vacuum forming, the sheet is clamped round its circumference and heated. A vacuum is applied from below to draw the sheet into a female die. On cooling, the plastic sheet acquires the form of the die.

# Exercises XX

**1.**

a) Automatic lathes are used for the production of a large number of identical components.

b) In the past, small batches of components were produced by the use of conventional machines and skilled labour.

c) The disadvantages of the previous methods of production, were the low machine utilization time, and the high cost of production.

d) Nonmachining time is reduced by using automatic tool changing, automatic tool resetting, and the automatic changing of workpieces.

e) CNC machines have drives which are controlled by a computer, unlike ordinary machines which have drives controlled by a skilled operator.

f) The machine utilization time is high, because the waiting time required for resetting the machine is small.

g) Only a change in the computer program is necessary when a new batch of components has to be produced.

h) The operator's skill is unimportant, because the accuracy of the work is only dependent on the accuracy of the machine and the program.

i) Each tool is held in an adaptor and stored in a magazine. The required tool is selected from a magazine and used to replace the one in the machine.

j) Minimization of wear is achieved by using rolling components, instead of sliding components.

**2.**

a) Automatic lathes are used for the production of a large number of identical components.

b) A new development has been the use of numerically controlled machines for small batch production.

c) Previously, small batch production was achieved using skilled labour and conventional machine tools.

d) This resulted in a low machine utilization time and a high cost of production.

e) The operation of a CNC machine is controlled by a computer program.

f) Special tools like jigs and fixtures are not required when a CNC machine is used.

g) Minimization of wear in CNC machines is achieved by using rolling components instead of sliding components.

h) Measuring instruments give a continuous indication of the position of the cutting tool.

i) The correct tool for a particular operation is selected automatically from a magazine.

j) The massive construction of the machines gives them the stability needed to withstand large cutting forces.

**3.**

a) Single and multispindle automatic lathes have found their greatest application, where a large number of identical components have to be produced. A new development has been the use of computer controlled machines, for the production of small batches of components.

b) In the past, small and large batches were produced on conventional machines, using highly skilled labour. This resulted in a low machine utilization time, a long waiting time, and high production costs.

c) The operation of a CNC machine tool is controlled by a computer. Machine tools which are used in this type of work, do not have the normal movements and drives used by conventional machines.

d) CNC machines do not need to be set up by a highly skilled operator. Only a change in the computer program is necessary to produce a new batch of components. The flexibility of the system results in high machine utilization times.

## Exercises XXI

**1.**

a) In an internal combustion engine the fuel is burnt inside the engine, while in an external combustion engine the fuel is burnt outside the engine.

b) The chemical energy of the fuel is converted into heat energy, and part of the heat energy is converted into mechanical energy by the engine.

c) An Otto cycle petrol engine has four strokes.

d) The connecting rod and the crankshaft have the function of changing the reciprocating motion of the piston into the rotatory motion of the crankshaft.

e) The power stroke is the only stroke in which the engine extracts energy from the fuel.

f) The four strokes in the Otto cycle are called induction, compression, power (or ignition), and exhaust.

g) In the induction stroke, the inlet valve opens and the piston moves downward. This creates a partial vacuum that sucks in the petrol-air mixture into the cylinder through the open inlet valve.

h) The flywheel stores the energy generated during the power stroke, and uses this energy to move the crankshaft, connecting rod, and pistons, through the other three strokes.

i) The chief difference between the two engines lies in the different methods used for introducing fuel into the cylinder and for burning fuel in the cylinder.

j) Multicylinder engines run more smoothly and silently than single cylinder engines.

**2.**

a) In an internal combustion engine the fuel is burnt inside the engine.

b) The chemical energy of the fuel is converted into heat energy by combustion, and part of this is converted into mechanical energy by the engine.

c) The valves can be opened or closed to allow a petrol-air mixture to enter or leave the cylinder.

d) The connecting rod and the crankshaft enable the reciprocating motion of the piston to be converted into rotatory motion.

e) During the power stroke, the engine extracts energy from the fuel.

f) During the induction stroke, the inlet valve opens and the piston moves downwards.

g) During the compression stroke both valves remain closed and the piston moves upwards compressing the petrol-air mixture.

h) The sprocket wheel drives the camshaft through a chain drive.

i) The difference between a petrol and a diesel engine lies in the different methods used for introducing and burning fuel in the cylinder.

j) Diesel engines have a higher thermal efficiency than petrol engines and therefore consume less fuel.

**3.**

a) The petrol engine and the diesel engine, are both combustion engines in which the fuel is burnt inside the engine.The chemical energy of the fuel is first converted into heat energy by combustion, and then into mechanical energy by the engine.

b) Most petrol engines are based on the Otto four stroke principle, in which a mixture of petrol and air is first compressed, and then ignited.

c) The petrol engine has a cylinder with a tight fitting piston, which can undergo an up and down motion. In the head of the cylinder, there is an inlet and an outlet valve. These valves can be opened or closed to allow a petrol-air mixture to enter or leave the cylinder.

d) In a four stroke process, the movement of the piston takes place in four stages, two upwards and two downwards. These four strokes are called induction, compression, ignition, and exhaust. Energy is extracted from the fuel, only during the ignition stroke.

e) The petrol engine and the diesel engine are similar, as far as their mechanical components are concerned. The main difference between the engines, lies in the methods used for introducing the fuel into the engine, and for burning the fuel.

# Exercises XXII

## 1.

a) Market studies of various types are used to identify the kinds of products that can be sold at a profit.

b) A well designed product should be visually appealing, and be able to perform its function satisfactorily and reliably. It should be easy to use, easy to maintain, and easy to manufacture at a competitive price.

c) Products should be made in such a way that they can be easily recycled or disposed of at the end of their useful life without damaging the environment.

d) Outsourcing refers to the purchase of components from outside suppliers.

e) A bill of materials is a list of all required components, materials, etc.

f) Components having similar features may be made more economically by using group technology.

g) Jigs are used to hold the workpiece and guide the tools during machining.

h) The processes involved in material requirements planning are the planning, scheduling, and controlling of production.

i) Such feedback is necessary for the improvement of old products, for the development of new products, and for improving the efficiency of the manufacturing process.

j) The layout of machines and other equipment should be optimized.

## 2.

a) In manufacturing enterprises the activities are arranged in a closed loop.

b) Product preferences may differ from country to country or from age group to age group.

c) Market studies identify products which can be manufactured profitably.

d) It may be cheaper to purchase mass produced components than to manufacture them.

e) Group technology is used to identify parts which have similar features.

f) Old processes have to be improved and new processes have to be developed.

g) The quality and dimensions of the components have to be systematically checked.

h) The final stage consists of the assembling of the components into the final product.

i) Programs for controlling product quality and for the maintenance of machines are essential.

j) The feedback of customer information closes the loop which coordinates the activities of a manufacturing enterprise.

3.

a) Market studies of various types are used to identify the kind of product which can be profitably manufactured and sold in a given environment.

b) A well designed product must be visually appealing and have the ability to perform its function satisfactorily and reliably through its expected life.

c) A further constraint imposed on manufactured products is an environmental one. Products should be made in such a way that they can be easily disposed of or recycled at the end of their useful life.

d) In the production planning stage, the most appropriate materials, processes, and process parameters have to be chosen. This is no easy task, given the large number of material and processes which are available today.

# Exercises XXIII

**1.**

a) The activities of a manufacturing enterprise are controlled today by a hierarchy of computers.

b) The modern approach to manufacturing is a system approach which is different from the traditional approach which regarded manufacturing as a sequence of individual steps.

c) A CAD system consists of a computer equipped with suitable software.

d) The data generated in a CAD process is stored in a common data base.

e) An increase in efficiency is achieved by combining CAD and CAM into a common process.

f) Material requirements planning is used for the control of inventories, and for the timing of the requirements of materials, components, subassemblies, etc.

g) The term releasing refers to the process of initiating production orders.

h) The automatization of the entire manufacturing process from design to finished product is called CIM.

i) In the scheduling process, each order is assigned to a sequence of machines.

j) The term inventory on hand refers to the items in stock or being produced.

**2.**

a) The traditional approach to manufacturing was to view it as a sequence of individual steps.

b) Detailed product design was done by making drawings on a drawing board.

c) The modern design method which is called computer aided design uses a computer with suitable software.

d) The object which is being designed can be viewed on a video display terminal.

e) The proper use of computers facilitates the manufacturing process, and increases the competitiveness of an enterprise.

f) Releasing is the process of initiating production orders.

g) These requirements are compared with the inventory on hand.

h) An increase in efficiency can be achieved by combining CAD and CAM.

i) Lead time is the time taken between the placing of an order and the delivery of the product.

j) Reporting is the process of informing management about how the manufacturing process is proceeding.

**3.**

a) Competitive pressures have exposed the shortcomings of this view, and compelled manufacturers to adopt a system approach, in which the various activities form part of an interacting dynamic system.

b) Detailed product design was done in the past by making drawings on a drawing board. The modern method which is called computer aided design uses a computer with suitable software.

c) By using geometric modelling and analysis, different design variations can be studied. A particular advantage of this method is that the object can be viewed from different angles on the screen.

d) An increase in efficiency can be realized by integrating CAD and CAM into a common CAD/CAM process. A common data base is necessary to enable the entire process from the design stage to the manufacturing stage to be automated.

# Exercises XXIV

**1.**

a) Productivity is usually defined as the value of goods produced per employee.

b) Direct costs refer to the actual cost of manufacturing.

c) The term net material cost refers to the cost of the raw material minus the cost of the remaining scrap material.

d) Energy costs that are not directly used in production are included in the indirect costs.

e) The cost savings due to any improvement in manufacturing productivity may be nullified if indirect costs are not carefully controlled.

f) Fixed costs include the costs of buildings, equipment, and all other facilities in general.

g) The break even number is the minimum number of units that will have to be produced before a profit can be made.

h) Direct costs are proportional to the number of units produced.

i) The term outsourcing refers to the purchase of components, subassemblies, etc. from outside suppliers.

j) The manufacture of all the components that they use, would require more capital, more expertize, and more production facilities than their resources would allow.

**2.**

a) The ultimate criterion of the value of a product is its selling price.

b) Productivity is usually defined as the value of goods produced per employee.

c) The total cost can be divided into direct costs, indirect costs, and fixed costs.

d) Direct costs are proportional to the number of units produced

e) The division of costs into direct and indirect costs is a fuzzy one.

f) The improvement in productivity may be nullified if indirect costs are not carefully controlled.

g) The fixed cost per unit is halved if a machine is used for two shifts a day.

h) Specialized suppliers can produce a component of higher quality cheaper than the manufacturer himself.

i) Automobile manufacturers buy a large percentage of their components from outside suppliers.

j) What is important is that all costs be included in the cost calculations.

3.

a) The unit cost of a product can be estimated without much difficulty. Productivity is however a much more difficult quantity to measure. It is usually defined as the value of production per employee.

b) Automation calls for heavy investment in machines, and is only profitable if large quantities of goods are produced. Unit costs are reduced by automation, but an economic limit is reached after which further investment yields diminishing reductions in unit costs.

c) Direct costs refer to the actual cost of manufacturing, and are proportional to the number of units produced. Such costs can be divided into two parts, material costs and labour costs.

d) The raw material may be in many forms like sheet metal, powder material, or even in the form of castings or forgings, possibly obtained from an outside supplier. Labour costs refer to the cost of the personnel employed. Both these costs can be calculated per unit produced.

e) Every manufacturer has to consider whether it is not better to buy a component or product from an outside supplier who can produce a component of higher quaity cheaper than the manufacturer himself.

# Introduction to part 2

In the first part of this book the average length of the text contained in each chapter was about two pages. In addition to this, the chapters contain drawings, vocabularies, exercises and solutions. The texts have been deliberately made short in order to allow room for the exercises and solutions. Although the exercises and solutions are necessary to help in the understanding of the text, the primary source of information about the subject remains the text. It is now felt that the texts are too short and that it would be desirable to make the texts longer in future.

In the sixth edition, the book has been divided into two parts. The first part contains the 24 chapters which already existed in unchanged form. The second part contains texts each about four pages in length, with drawings and a vocabulary, but without exercises or solutions. In this way it is hoped that more information relating to engineering themes and more new technical words can be introduced using a smaller number of pages.

Since the students have already gained experience in doing exercises, it is hoped that they will now be in a better position to absorb the material in the texts without having to do any more exercises. If this book is being used as the basis of a taught course in technical English, the teacher can easily make up suitable exercises if and when required.

## Text 1  Basics of robots

Many definitions of robots have been provided to distinguish them from other forms of automation. One such definition is as follows:

"A robot is a computer controlled mechanical device which can be programmed to carry out a variety of tasks automatically without human supervision".

Many remotely controlled *mechanical manipulators* are often called robots, (for example those that are used to *handle nuclear materials*) although they are not robots. They should only be called robots, if they remember their instructions and repeat their tasks automatically. A robot has a mechanical system and a control system which usually includes a computer.

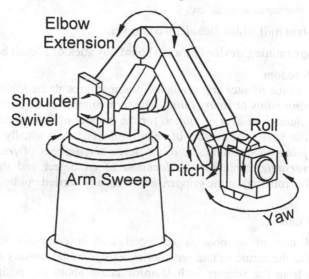

Fig 1 Robot with six rotary joints including three wrist joints

**Components of a typical robot**

1. **A base** which can be fixed or movable.

2. **An arm assembly** (also called a **manipulator**), which has the following components:

    (a) **Links** which are the rigid parts of an arm.

    (b) **Joints** which provide a *movable connection* between the links. Two types of joints are normally used, **sliding joints** and **rotary joints**.

    (c) **Actuators** which are the drive systems that provide power for the movements of the different components of a robot. Three types of actuators are commonly used, **electromechanical**, **hydraulic** or **pneumatic**.

(d) **The wrist** is the name given to the last three joints on the robot's
arm. They are rotary joints and their axes of rotation are mutually
perpendicular. The rotary movements are called **yaw** (vertical axis),
**pitch** (horizontal axis from left to right), **roll** (horizontal axis from
back to front through the wrist) (see Fig 1).

(e) **The end-effector (**or end of arm tooling) comes after the wrist, and
is the last part of the arm. End effectors can be classified into two
groups (1) **grippers** which are used to grip and move objects and
(2) **specialized tools** which are used for special jobs.

(f) **Sensors (or encoders)** are used to track the motion of the joints on
the manipulator. Sensors can be of various types, electronic,
magnetic, optical, etc.

3. **A control unit** which includes a computer.

4. **A programming device** like a keyboard, joy stick or a teach box.

**Degrees of freedom**

The motions of the roboter arm are made possible through the joints which can
be either *sliding joints* or *rotary joints*. The term number of degrees of freedom
refers to the number of axes of motion (or the number of independent motions)
that an arm can have. The number of degrees of freedom is usually equal to the
*number of joints* in the robot. A robot needs *six degrees* of *freedom* to be
*completely versatile*, three for the *location* of an object and three for the
*orientation* (or rotation) of the gripper, a task usually carried out by the motions
of the wrist.

**The control unit**

The control unit of a robot is equipped with a computer which *sends
instructions* to the actuators and *receives signals* from the *sensors* of the robot.
The signals from the sensors include information about the positions of the
*robot's joints*, information about the robot's *gripper* (whether open or closed),
information from the *vision sensor*, and also information from the *touch sensor*
on the gripper. The controller *interprets* the *received information* and transmits
*additional signals* to ensure that the current task is completed as planned.

**Robot control systems**

Two types of control systems are commonly used, *closed loop* and *open loop*. In
a *closed loop control system*, the controller transmits an instruction to an
actuator to move a joint, and a sensor on the joint sends a report to the
controller, indicating how the joint has moved in response to the instruction sent
by the controller. A closed loop is thus formed from the controller to the
actuator to the sensor and back to the controller (see Fig 2). If the joint did not
move as instructed, the controller sends additional *correction signals* until the
joint reaches the required position or *end point*. A robot with a closed loop

control system is usually called a ***servo robot***. A servo robot can control the ***velocity***, ***acceleration*** and ***path of motion***, in addition to the ***end points***.

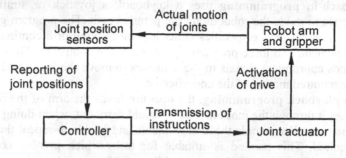

Fig 2 Closed loop control system

In an ***open loop control system***, there are no sensors which measure how the limbs move in response to the instructions sent by the controller. What is known is the end position, but not whether the limb has reached this position. Such robots which are also called ***non servo robots*** or ***pick-and-place robots***, are only controlled at their end points but not along their paths.

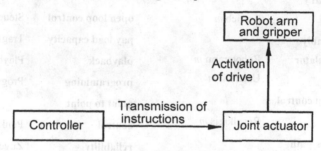

Fig 3 Open loop control system

## The movement of a robot

In robot terminology, the word ***trajectory*** refers to the path of the end effector. There are in general two types of trajectory, point to point (PTP) and continuous path (CP). In PTP, only the end points of the motion are important. In CP motion, both the ***connecting path*** as well as the ***end points*** are important.

Robots that do material handling, machine tool loading or spot welding could use either PTP or CP motion. Robots that do ***spray painting*** could only use CP motion to perform their work satisfactorily. Two terms that are important when defining trajectories are:

1. **Accuracy** which is a measure of how closely the robot arm reaches the required position.
2. **Repeatability** which refers to the ability of a robot arm to return to the same position.

**The programming of a servo robot**
Three methods are commonly used to program servo robots.
1. **Teach-in programming** uses a keyboard, a joystick or similar input device to guide the robot along the planned path. The program generated by the input device specifies either a point to point or a continuous path movement, and these program signals go to the controller. The controller sends appropriate signals to the actuators to move the joints and produce the required motion of the end-effector.
2. **In playback programming**, the operator holds the arm of the robot and takes it through the motions that it should carry out, when doing the task. The controller records these movements and is able to repeat them when required. This method is suitable for tasks which involve continuous motion like spray painting, arc welding, cutting, etc. A program obtained on one robot can be used on other robots to carry out the same task.
3. **Off-line programming** is possible in a manner similar to that used to program CNC machines. This is done *without involving* the *robot* itself as in the previous two methods.

## Vocabulary

| | | | |
|---|---|---|---|
| **actuator** | Antrieb *m* | **open loop control** | Steuern *v* |
| **arm assembly** | | **pay load capacity** | Tragfähigkeit *f* |
| **or manipulator** | Roboterarm *m* | **playback** | Playback |
| **base** | Unterteil *m* | **programming** | Programmierung *f* |
| **closed loop control** | | **point to point** | |
| **system** | Regelkreis *m* | **control** | Punktsteuerung *f* |
| **continuous path** | | **reliability** | Zuverlässigkeit *f* |
| **control** | Bahnsteuerung *f* | **repeatability** | Wiederhol-genauigkeit *f* |
| **degrees of freedom** | Freiheitsgrad *m* | **resolution** | Auflösung *f* |
| **end effector** | Arbeitsorgan *n* | **teach-in** | Teach-in |
| **end point** | Sollposition *f* | **programming** | Programmierung *f* |
| **gripper** | Greifer *m* | **trajectory** | Trajektorie *f* |
| **joint** | Gelenk *n* | **work zone or** | |
| **link** | Glied *n* | **work envelope** | Arbeitsraum *m* |
| **off-line** | Indirect oder off-line | **wrist** | Handgelenk *n* |
| **programming** | Programmierung *f* | | |
| **orientation** | Orientierung *f* | | |

# Text 2  A visit to an Industrial Fair

## 1.  A visit to the stand of a CNC machine manufacturer

Two South American engineers Carlos Santana and Gerry Gomez are visiting the Hannover Fair in Germany to investigate the possibilities of buying some machines for their manufacturing plants at home. Their first visit is to the firm Hugo Meyer AG, a major manufacturer of CNC machines. They first go to the information office to get a map showing the location of the firm's stand.

Santana:  Hello, good morning. Would it be possible for us to get a map which shows the locations of the stands of the exhibiting firms.

Assistant:  Here you are sir. I can give you a free copy of our exhibition map.

Santana:  Thank you.

The engineers are able to find their way to the stand quite quickly. At the stand they are greeted by the sales manager Hans Berger.

Berger:  Good morning, gentlemen. Welcome to our stand. I am Hans Berger the sales manager. My staff and I are ready to help you in every way.

Gomez:  I am Gerry Gomez and this is my friend Carlos Santana. We have come from South America to buy some machinery for use in our plants.

Berger:  Our firm specializes in the manufacture of CNC machines. We have a wide range of machines here, and you are welcome to inspect them under working conditions both here and in our factories.

Santana:  Can you tell us more about the machines that you manufacture. -

Berger:  First of all, we have the standard machines like boring mills, machining centres and turning centres, both vertical and horizontal. These are extremely versatile and reliable machines sold at very attractive prices.

Secondly we have more sophisticated systems of machines which considerably improve productivity. For example we have a multiple turret two station machining centre for five axis working, and a modular multi-station machining centre for four axis working.

Other systems that we can offer are transfer lines where individual machining sequences are performed by separate CNC machines. We also manufacture hybrid systems consisting of processing centres and transfer lines which achieve an optimum price to performance ratio. All machines and systems including the standard machines can be tailor-made to suit your requirements.

Gomez:  What about the dimensions and weights of the workpieces that each machine can handle. The machines must be of the right size to handle our workpieces in the most cost-effective fashion.

Berger:  Each type of machine that we manufacture is made in a range of sizes. For example in our silver star series, customers can choose

|          | machining centres with pallet sizes ranging from 400 mm x 500 mm to 1250 mm x 1600 mm and loads of between 600 kg and 6800 kg. |
|----------|------|
| Santana: | We have brought some samples of workpieces that are normally machined in our plants at home. We would like to see these being machined on your CNC machines. This will give us an indication of the time taken for loading, machining, etc. and also show us how good the surface finish will be. |
| Berger:  | We will be very happy to do this, Sir. Our engineers will show you the whole procedure, starting from the writing of the program, through the loading and machining processes to the final product. |
| Gomez:   | We have also brought with us data concerning the range and quantities of items that we produce, and in addition lists of existing machines and floor layout plans of our present production facilities. We would like your advice on what machines would be best suited for our work, and also your concepts of the best floor layout plan and combinations of machines which will optimize our production. Where possible we would like to continue using some of our existing machines, which are mostly in good condition. |
| Berger:  | Our systems engineers will help you to choose the machines that are best suited for your use, and work out the best layout for the optimization of your production. |
| Santana: | At this point we must mention that our financial resources are limited, and that we must work within these limitations. |
| Berger:  | Of course, Sir. This is one of the first things that we consider. We try to do the best that we can for our customers, while remaining within the financial limits that are imposed on them. |
| Gomez:   | What about servicing, repairs and maintenance of the machines. |
| Berger:  | We have training courses on the operation, maintenance and repairs of our machines for the technicians sent by our customers. In addition we have agents for our machines in your country, who are highly experienced in repair and maintenance. You can depend on them for help when you run into difficulties. We have also comprehensive manuals which give descriptions of our machines, together with detailed instructions for maintenance and repairs. |
| Santana: | Can you help in obtaining finance for the purchase of the machines ? |
| Berger:  | We certainly can. The financial terms are among the most crucial factors in transactions of this nature. We help our customers to obtain the necessary credit on easy terms from a suitable bank. This allows them to stretch payments over a period of several years. |
| Gomez:   | It is clear that we have a lot of things to discuss. I suggest that we spend a few hours every morning for the next few days, considering in detail, all the points that we have mentioned so far. |

## 2. A visit to the stand of a firm making solar cells and sensors

Santana: Here is the next firm that we wanted to visit, Gerry. Sensors and Solar Cells Inc.

Engineer: Welcome, Sir. I am Frank Kohler, sales engineer at your service.

Gomez: We are engineers from South America and would like to know details about your solar cells and also your large solar modules.

Engineer: Our solar cells are entirely made from crystalline silicon. These cells have a theoretical efficiency of about 30 % but in practice our solar cells achieve an efficiency of only about 15 % . The cost of the generated electricity is about three or four times higher than the cost of the electricity delivered by conventional power stations.

Santana: What limits the theoretical efficiency of the solar cells ?

Engineer: The cells are only able to generate electricity from a narrow band of the solar spectrum, depending on the electronic bandgap of the material. Research is being carried out on solar cells which combine a number of materials having different bandgaps. These multiple bandgap cells are said to have theoretical efficiencies of up to 87 % .

Gomez: We would also like to know about the sensors that you manufacture.

Engineer: We have a wide range of sensors both active and passive which can detect and measure a variety of physical quantities. Their resolution, accuracy and sensitivity are specified in our catalogue. We can also make special types for you to order. These will naturally have longer delivery times and cost more money.

Santana: Can you tell us more about the types of sensors that you make.

Engineer: We have sensors for the measurement of temperature and pressure, sensors for path and angle measurement, sensors for the measurement of force and strain, proximity sensors and optical sensors.

Gomez: How do you increase the sensitivity of the sensors, and is the output in analogue or digital form ?

Engineer: Our sensor units have built-in amplifiers and linearising elements which improve their sensitivity and accuracy. We can supply units with either digital or analogue output.

Santana : What about price, delivery dates and guarantees ?

Engineer: The price will depend on the quantities ordered. Most of the units are in stock and can be sent to you immediately. We guarantee our products for a period of two years provided they have been used in the correct way. Here is our catalogue, Sir. We would be pleased to show you all our products whenever you have the time to see them. You are also welcome to visit our factory by appointment.

Gomez: We will go through your catalogue and decide what to do. Bye-bye.

Engineer: Bye-bye, gentlemen. We will be looking forward to your next visit.

### 3. A visit to the stand of the firm Advanced Robotics Inc.

Engineer:   Hello gentlemen, welcome. I am sales engineer Guido Schmidt.

Santana:    We are engineers from South America and would like to have information about the types of robots that you manufacture.

Engineer:   We manufacture a range of robots of both the servo and nonservo types, like for example pick and place robots, gantry robots, SCARA robots, and jointed arm (or revolute) robots. We can also make special robots designed to meet your individual requirements.

Our robots are flexible automation tools and our customers have bought and used them to do many kinds of industrial work.

Santana:    Can you give us more details about some of these applications.

Engineer:   Our robots have been used for material handling, machining, casting, forging, welding, finishing, assembling, inspection, etc. and also for continuous processes like those in chemical plants.

Gomez:      How do you achieve such a high level of flexibility ?

Engineer:   First of all we have many different types of robots. Secondly, a robot can be programmed repeatedly to do different tasks. Thirdly, a variety of different kinds of support equipment can be used to help the robot. For example one can choose from many types of grippers and end of arm tooling. Sensors can be used to help in many applications like controlling the movements of the manipulator, doing assembly or material handling operations, locating objects, and inspecting objects for flaws or defects.

Santana:    How can you be sure that you have chosen the most suitable type of robot for a particular application ?

Engineer:   There are a number of simulation tools available, which enable the user to evaluate the performance of different robots when performing a particular task, without having to purchase them. Such tools can also help in optimizing the factory layout and developing applications without actually using a robot.

Gomez:      What about the use of teams of robots on the factory floor.

Engineer:   Using several robots is a particularly challenging assignment, and our experience helps us to find the optimum solution in each case. In most applications, it is desirable that simulation tools be used to compare the different solutions that are possible.

Santana:    We think that the productivity in our factories at home can be improved by the use of robots. We would like to discuss with you in detail the best way of solving some of our manufacturing problems

Engineer:   The best person to advise you will be our systems engineer Uwe Schneider. If you come with me, I will take you to him.

The two South Americans follow him to meet the systems engineer Uwe Schneider and discuss their problems with him.

# Vocabulary

| | | | |
|---|---|---|---|
| accuracy | Genauigkeit *f* | modular system | Baukasten-system *n* |
| application | Anwendung *f* | module | Baukasten *m* |
| appointment | Verabredung *f* | performance | Leistung *f* |
| assignment | Aufgabe *f* | physical quantity | physikalische Größe |
| attractive price | verlockende Preis *m* | pick and place robots | Pick and place Roboter *m* |
| challenging | herausfordernd *adj* | plant | Anlage *f* |
| comprehensive | umfassend *adj* | price to performance ratio | Preis-Leistungs-verhältnis *n* |
| conventional | herkömmlich *adj* | procedure | Verfahren *n*, Vorgehen *n* |
| cost-effective | kostengünstig *adj* | proximity sensors | Näherungs-sensoren *pl* |
| crucial | kritisch *adj* | range | Auswahl *f*, Angebot *n* |
| deliver | liefern *v* | ratio | Verhältnis *n* |
| delivery time | Lieferzeit *f* | reliable | zuverlässig *adj* |
| desirable | wünschenswert *adj* | repeated | wiederholt *adj* |
| details | Einzelheiten *pl* | resolution | Auflösung *f* |
| efficiency | Wirkungs-grad *m* | sensitivity | Sensitivität *f*, Feingefühl *n* |
| evaluate | bewerten, berechnen *v* | sequence | Reihenfolge *f* |
| financial limits | Kostengrenze *f* | servicing | Dienstleistung *f* Wartung *f* |
| financial resources | Geldmittel *pl* | solve | lösen *v* |
| gantry robots | Portalroboter | sophisticated | hochentwickelt, raffiniert *adj* |
| generate | erzeugen *v* | strain | Verformung *f* |
| indication | Andeutung *f*, Angabe *f* | surface finish | Oberflächen-beschaffenheit *f* |
| investigate | ermitteln, untersuchen *v* | tailor-made | nach Maß angefertigt |
| layout | Anordnung *f* | transfer line | Förderanlage *f* |
| loading | Ladung *f* | turning centre | CNC Dreh-maschine *f* |
| machining centre | Bearbeitungs-zentrum *n* | turret | Revolverkopf *m* |
| maintenance | Instandhaltung *f* | versatile | vielseitig *adj* |
| manipulator | Roboterarm *m* | wide range | Große Auswahl *f* |
| material handling techniques | Handhabungs-technik *f* | | |

## Vocabulary 1   Englisch / Deutsch

| | |
|---|---|
| ability | Fähigkeit *f* |
| abrasion | Abrieb *m* |
| abrasive | Schleifmittel *n* |
| abrasive wheel | Schleifscheibe *f* |
| absolutely | vollkommen *adj* |
| accessory | Zubehör *n* |
| accomplish | ausführen *v* |
| accordance | Übereinstimmung *f* |
| accounting | Buchführung *f* |
| accuracy | Genauigkeit *f* |
| accurate | genau *adj* |
| achieve | leisten *v* |
| acid | Säure *f* |
| acquire | erwerben *v* |
| actuator | Antrieb *m* |
| adaptor | Aufnehmer *m* |
| administration | Verwaltung *f* |
| advance | vorwärtsbewegen *v* |
| advanced | fortgeschritten *adj* |
| advantage | Vorteil *m* |
| alignment | Ausrichtung *f*, Anordnung *f* |
| allowance | Spiel *n* |
| alloy | Legierung *f* |
| although | obwohl *cj* |
| amongst | unter *pr* |
| amount | Menge *f* |
| anneal | weichglühen *v* |
| anticlockwise | entgegen dem Uhrzeigersinn |
| appeal | appellieren *v* |
| application | Anwendung *f* |

| | |
|---|---|
| appointment | Verabredung *f* |
| appraise | schätzen *v* |
| approach | Methode *f*, Zugang *m* |
| appropriate | passend, geeignet *adj* |
| arbor | Fräserdorn *m* |
| arc | Lichtbogen *m* |
| arm assembly or manipulator | Roboterarm *m* |
| ascend | ansteigen *v* |
| assemble | montieren *v* |
| assess | bewerten *v* |
| assignment | Aufgabe *f* |
| associated | verbunden *v* |
| attachment | Zusatzgerät *n* |
| attractive | anziehend *adj* |
| automatic | automatisch *adj* |
| available | vorhanden *adj* |
| awkward | schwer zugänglich *adj* |
| axis | Achse *f* |
| backlash | Flankenspiel *n* |
| balance | Gleichgewicht *n* |
| bar | Stange *f*, Stab *m* |
| barrel finish | trommelpolieren *v* |
| base | Unterteil *m* |
| based on | begrundet auf *v* |
| basic size | Nennmass *m* |
| batch | Menge *f* |
| bead | bördeln, falzen *v* |
| bearing | Lager *n* |
| bed | Bett *n* |

| behaviour | Verhalten *n* |
| belt | Riemen *m* |
| bend | biegen *v* |
| bevel gear | Kegelrad *n* |
| bilateral | zweiseitig *adj* |
| blacksmith | Schmied *m* |
| blank | stanzen *v* |
| bolt | Schraube *f* |
| | (mit Mutter) |
| bolt and nut | Schraube und |
| | Mutter *f* |
| bonding | Bindemittel *n* |
| material | |
| braze | hartlöten *v* |
| break-even | Rentabilitätsgrenze *f* |
| number | |
| breaking | Bruchfestigkeit *f* |
| strength | |
| brittle | spröde *adj* |
| brittleness | sprödigkeit *f* |
| broach | räumen *v* |
| buff | hochglanzpolieren *v* |
| bulk deform | massivumformen *v* |
| burn | verbrennen *v* |
| bush | Lagerbuchse *f* |
| camshaft | Nockenwelle *f* |
| capstan lathe | Revolverdrehbank *f* |
| carbon steel | Kohlenstoffstahl *m* |
| carburettor | Vergaser *m* |
| carburize | aufkohlen *v* |
| carriage | Werkzeugschlitten *m* |
| carry out | durchführen *v* |
| case harden | einsatzhärten *v* |
| cast iron | Gusseisen *n* |
| casting | Guss *m* |

| cavity | Hohlraum *m* |
| centreless | Spitzenlos |
| grinding | schleifen *v* |
| chain | Kette *f* |
| chamber | Kammer *f* |
| change | Änderung *f* |
| change | wechseln *v* |
| character | Eigenart *f* |
| cheap | billig *adj* |
| check | prüfen *v* |
| chisel | Meißel *m* |
| choose | wählen, aufsuchen *v* |
| chuck | Futter *n* |
| circumference | Umkreis *m* |
| clamp | Klemme *f* |
| clean | reinigen *v* |
| clearance | Spielraum *m*, Spiel *n* |
| clearance angle | Freiwinkel *m* |
| clockwise | im Uhrzeigersinn |
| closed loop | geschlossener |
| | Kreis *m* |
| closed loop | Regelkreis *m* |
| control system | |
| clutch | schaltbare |
| | Kupplung *f* |
| coarse grain | grobkörnig *adj* |
| collar | Muffe *f* |
| column | Säule *f* |
| combustion | Verbrennung *f* |
| compel | zwingen, nötigen *v* |
| compensate | kompensieren *v* |
| competition | Konkurrenz *f* |
| complex | komplex *adj* |
| composed of | zusammengesetz |
| | auf *v* |

| | | | |
|---|---|---|---|
| composite | zusammengesetz *adj* | counterbore | zylindrische |
| compound | Verbindung *f* | | Senkung *f* |
| compound | Recktecktisch *m* | countersink | Kegelsenkung *f* |
| table | | coupling | Kupplung *f* |
| comprehensive | umfassend *adj* | crack | Riss *m*, Spalt *m* |
| compress | zusammendrücken *v* | crank | Kurbel *f* |
| conceive | vorstellen *v* | crankshaft | Kurbelwelle *f* |
| concerned | betroffen *v* | create | erzeugen *v* |
| cone | Kegel *m* | creep | kriechen *v* |
| conical | kegelförmig adj | criterion | Maßstab *m* |
| connecting rod | Pleuelstange *f* | cross-section | Querschnitt *m* |
| consider | erwägen, | crystal grain | Mikrokristall *n* |
| | überlegen *v* | customer | Kunde *m* |
| consist of | bestehen aus *v* | cutting edge | Schneidkante *f* |
| constraint | Zwang *m*, | decrease | sich vermindern *v* |
| | Nötigung *f* | deep draw | tiefziehen *v* |
| construction | Anlage *f* | deformation | Verformung *f* |
| contain | enthalten *v* | degrees of | Freiheitsgrad *m* |
| content | Inhalt *m* | freedom | |
| continuous | ständig *adj* | delivery | Lieferung *f* |
| continuous | Bahnsteurung *f* | demand | Bedarf *m* |
| path control | | department | Abteilung *f* |
| conventional | herkommlich *adj* | depend | abhängen *v* |
| conversion | Umsetzung *f* | depth | Tiefe *f* |
| convert | umwandeln *v* | description | Beschreibung *f* |
| coolant | Kühlschmiermittel *n* | design | entwerfen *v* |
| coordinate | anordnen *v* | determine | bestimmen, |
| copper | Kupfer *n* | | entscheiden *v* |
| corresponding | entsprechend *adj* | development | Entwicklung *f* |
| corrosive | korrosiv, | deviation | Abweichung *f*, |
| | ätzbar *adj* | | Abmaß *n* |
| costs | Kosten *pl* | device | Gerät *n* |
| cost-effective | kostengünstig *adj* | dial gauge | Messuhr *f* |
| cotter | Keil *m* | die | Schneideisen *n* |
| cotter pin | Splint *m* | die block | Schneidplatte *f* |

| die casting | Spritzguss *m* | end effector | Arbeitsorgan *n* |
| die set | Schneidwerkzeug *n* | end point | Sollposition *f* |
| difficult | schwer *adj* | energy | Energie *f* |
| dimension | Maß *n* | engine | Motor *m* |
| diminish | vermindern, | engine head | Zylinderkopf *m* |
| | verringern *v* | engrave | gravieren *v* |
| direction | Richtung *f* | ensure | sicherstellen *v* |
| disadvantage | Nachteil *m* | enterprise | Unternehmen *n* |
| disengage | ausschalten *v* | environment | Umgebung *f* |
| dispose of | beseitigen *v* | equipment | Ausstattung *f*, |
| dividing head | Universalteilkopf *m* | | Einrichtung *f* |
| dowel pin | Dübel *m* | essential | wesentlich *adj* |
| downwards | abwärts *adv* | estimation | Schätzung *f* |
| drawing | Zeichnung *f* | evaluate | berechnen *v* |
| drift | Austreiber *m* | example | Beispiel *n* |
| drill | Bohrer *m* | execute | durchführen *v* |
| drive | treiben *v* | exhaust | Auspuff *m* |
| drop forging | Gesenkschmieden *n* | exist | existieren *v* |
| drum | Trommel *f* | experience | Erfahrung *f* |
| ductile | dehnbar, biegbar *adj* | expertise | Sachkenntnis *f*, |
| ductility | Dehnbarkeit *f* | | Fachkenntnis *f* |
| ecology | Ökologie *f* | extensive | umfassend *adj* |
| economical | sparsam *adj* | extent | Bereich *m* |
| efficiency | Wirkungsgrad *m*, | external | äußerlich *adj* |
| | Leistungsfähigkeit *f* | extract | ausschneiden *v* |
| efficient | leistungsfähig *adj* | extremely | sehr *adv* |
| eject | auswerfen *v* | extrude | strangpressen *v* |
| elastic limit | elastische Grenze *f* | facilitate | erleichtern, |
| element | Grundbestandteil *m* | | ermöglichen *v* |
| eliminate | beseitigen *v* | facility | Einrichtung *f*, |
| elongate | verlängern *v* | | Anlage *f* |
| elongation | Verlängerung *f* | fatigue strength | Dauerfestigkeit *f* |
| emboss | prägen *v* | feature | Eigenschaft *f* |
| employ | beschäftigen *v* | feedback | Rückkopplung *f* |
| enable | ermöglichen *v* | ferrous | eisenhaltig *adj* |

| | | | |
|---|---|---|---|
| **filler** | Füller *m* | **gradually** | allmählich *adv* |
| **fine grain** | feinkörnig *adj* | **grain structure** | Gefüge *n* |
| **financial resources** | Geldmittel *pl* | **grains, granules** | Körnchen *n* |
| **finishing process** | Endverfahren *n* | **gravity die cast** | Gießen mit Schwerkraft *v* |
| **fit** | Passung *f* | **grease** | Fett *n* |
| **fixed costs** | Anlagekosten *pl* | **grind** | schleifen *v* |
| **fixture** | Vorrichtung *f* | **grip** | greifen *v* |
| **flange** | bördeln *v* | **groove** | Nut *f* |
| **flaw** | Defekt *m*, Fehler *m* | **guideways** | Bettführungen *pl* |
| **flexible** | beweglich *adj* | **half (pl. halves)** | Hälfte *f* |
| **flow** | fließen *v* | **hardwearing** | widerstandsfähig *adj* |
| **fluctuate** | schwanken *v* | **hardness** | Härte *f* |
| **fluid** | Flüssigkeit *f* | **headstock** | Spindlestock *m* |
| **flute** | Spannut *f* | **heat** | Hitze *f*, Wärme *f* |
| **flux** | Flussmittel *n* | **heat treatment** | Wärmebehandlung *f* |
| **flywheel** | Schwungrad *n* | **heavy duty** | Hochleistungs… *adj* |
| **follow-on die set** | Folgeschneidwerkzeug *n* | **helical gears** | Zahnräder mit Schrägverzahnung *pl* |
| **force** | Kraft *f* | **holder** | Behälter *m* |
| **force into** | einpressen *v* | **hollow** | hohl *adj* |
| **forecast** | voraussagen *v* | **hone** | honen *v* |
| **forge** | schmieden *v* | **hub** | Nabe *f* |
| **form** | formen, bilden *v* | **identical** | identisch *adj* |
| **frame** | Gestell *n* | **ignite** | entzünden *v* |
| **free-cutting steel** | Automatenstahl *m* | **impact** | Stoß *m*, Schlag *m* |
| | | **important** | wichtig *adj* |
| **freely** | zwanglos *adv* | **impose** | aufdrängen *v* |
| **friction** | Reibung *f* | **impossible** | unmöglich *adj* |
| **fuel** | Brennstoff *m* | **impression** | Prägung *f* |
| **fuel pump** | Benzinpumpe *f* | **improve** | verbessern *v* |
| **fuzzy** | undeutlich *adj* | **include** | einschließen, enthalten *v* |
| **gear wheel** | Zahnrad *n* | | |
| **generate** | erzeugen *v* | **increase** | zunehmen *v* |

| | | | |
|---|---|---|---|
| indentation | Eindruck *m* | linear | linear *adj* |
| indication | Anzeige *f* | link | Glied *n* |
| induction | Einführung *f* | load | Last *f* |
| ingot | Gussblock *m* | locking device | Losdrehsicherung *f* |
| injection nozzle | Einspritzdüse *f* | longitudinal | längs *adj* |
| inlet valve | Einlassventil *n* | loosen | lockern *v* |
| inspect | prüfen *v* | lubrication | Schmierung *f* |
| instruction | Befehl *m* | machine-hour | Machinenstundensatz- |
| interact | aufeinanderwirken *v* | rate | rechnung *f* |
| interchangeable | auswechselbar *adj* | machine tool | Werkzeugmaschine *f* |
| interfere | überlagern *v* | machining | Bearbeitungs- |
| interlock | ineinandergreifen *v* | allowance | zugabe *f* |
| internal | innerlich *adj* | machining | Bearbeitungs- |
| introduce | einführen *v* | centre | zentrum *n* |
| inventory | Bestands- | magnitude | Größenordnung *f* |
| | verzeichnis *n* | maintain | instandhalten *v* |
| invoice | Rechnung stellen *v* | maintainance | Instandhaltung *f* |
| involve | verwickeln *v* | manageable | beherrschbar *adj* |
| jig, fixture | Vorrichtung *f* | mandrel | Dorn *m* |
| join | verbinden *v* | mass produce | serien mäßig |
| joint | Gelenk *n* | | herstellen *v* |
| key | Keil *m*, Paßfeder *f* | material | Fördertechnik *f* |
| keyway | Nut *f* | handling | |
| knee | Konsole *f*, Knie *n* | material | deterministische |
| knowledge | Kenntnis *f*, Wissen *n* | requirements | Bedarfsermittlung *f* |
| labour costs | Arbeitskosten *pl* | planning(MRP) | |
| lack | mangeln, fehlen *v* | matt | glanzlos *adj* |
| lance | einschneiden *v* | measure | messen *v* |
| lap | läppen *v* | measurement | Messung *f* |
| lathe | Drehmaschine *f* | mechanism | Mechanismus *m* |
| layout | Anordnung *f* | mention | erwähnen *v* |
| lead | Blei *n* | metal chips | Metallspäne *pl* |
| lead | Steigung *f* | method | Methode *f* |
| leadscrew | Leitspindel *f* | micrometer | Messchraube *f* |
| limit | Grenze *f* | mild steel | Baustahl *m* |

| | | | |
|---|---|---|---|
| **minimum** | Minimum n | **practice** | ausführen v |
| **mixture** | Mischung f | **practice** | Gewohnheit f |
| **modular** | Baukasten- | **precision** | Genauigkeit f |
| **system** | System n | **predetermine** | vorbestimmen v |
| **module** | Baukasten m | **predictable** | voraussagbar adj |
| **molten metal** | flüssiges Metall n | **preference** | Bevorzugung f |
| **monitor** | überwachen v | **prepare** | vorbereiten v |
| **motion** | Bewegung f | **press** | drücken, pressen v |
| **mould** | formen, bilden v | **pressing** | gepresstes |
| **mould** | Gießform f, Form f | **(metal)** | Metallstück n |
| **movement** | Bewegung f | **pressure** | Druck m |
| **multipoint** | mehrschneidiges | **pressure** | Spritzguß m |
| **cutting tool** | Werkzeug n | **diecasting** | |
| **multipurpose** | mehrzweck- adj | **pressure pad** | Niederhalter m |
| **necessary** | notwendig adj | **prevent** | verhindern v |
| **necking** | Querschnitts- | **previous** | vorher adj |
| | verminderung f | **probably** | wahrscheinlich adv |
| **net** | netto adj | **procedure** | Vorgehen n |
| **nibble** | nibbeln, knabber- | **process** | Vorgang m |
| | schneiden v | **production** | Herstellung f |
| **noise** | Geräusch n, Lärm m | **progress** | Fortschritt m |
| **normalize** | normalglühen v | **progressive** | schrittweise adj |
| **notch** | Kerbe f | **project out** | vorspringen v |
| **nullify** | annullieren v | **projection** | vorspringender |
| **numerical** | numerische | **part** | Teil m |
| **control** | Steuerung f | **property** | Eigenschaft f |
| **open loop** | steuern v | **prototype** | Muster n |
| **control** | | **provide** | besorgen v |
| **performance** | Leistung f | **punch** | Stanzwerkzeug n |
| **pollution** | Umwelt- | | Schneidstempel m |
| | verschmutzung f | **punch** | stanzen, lochen v |
| **polycrystalline** | vielkristallin adj | **purchase** | kaufen v |
| **popular** | beliebt adj | **purpose** | Zweck m |
| **pour** | gießen v | **push** | schieben v |
| **power** | Leistung f | **quantity** | Größe f |

| | | | |
|---|---|---|---|
| quench | abschrecken *v* | report | Bericht erstatten *v* |
| quiet | ruhig *adj* | require | erfordern |
| rack | Zahnstange *f* | | brauchen *v* |
| rack and pinion | Zahnstangen- | resin | Harz *n* |
| | getriebe *n* | resistance | Widerstand *m* |
| rake angle | Spanwinkel *m* | resolution | Auflösung *f* |
| ram | Stößel *m* | reuse | wiederverwenden *v* |
| range | Reihe *f*, Bereich *m* | revert | zurückkehren *v* |
| rapid | schnell *adj* | revolution | Umdrehung *f* |
| ratio | Verhältnis *n* | revolve | sich drehen *v* |
| realize | verwirklichen *v* | ridge | Kamm *m* |
| reamer | Reibahle *f* | rigid | stabil *adj* |
| reciprocating | pendelnde | rigid coupling | starre Kupplung *f* |
| motion | Bewegung *f* | rigorous | streng, sorgfältig *adj* |
| recess | Einschnitt *m* | rivet | Niete *f* |
| recirculating | umlaufend *adj* | rod | Rundstab *m* |
| recycle | wiederbewerten *v* | rotary table | 360° drehbarer |
| reduce | vermindern, | | Schraubstock |
| | reduzieren *v* | rotary, rotatory | drehbeweglich *adj* |
| refer | verweisen, | rough | rau *adj* |
| | hinweisen *v* | route | weiterleiten *v* |
| refine | rückfeinen, | sale | Verkauf *m* |
| | vergüten *v* | schedule | Arbeitsterminplan *m* |
| regulate | regeln *v* | screw | Schraube *f* |
| reinforce | verstärken *v* | screw feed | Schnecken- |
| reinforced | Eisenbeton *m* | mechanism | strangpresse *f* |
| concrete | | screw thread | Gewinde *n* |
| reject | ausschneiden *v* | seamless | nahtlos *adj* |
| release | Auftrag erteilen *v* | select | auswählen *v* |
| reliability | Zuverlässigkeit *v* | self-centering | Dreibacken- |
| remain | übrigbleiben *v* | chuck | bohrfutter *n* |
| remove | entfernen *v* | semi-skilled | angelernt *adj* |
| repeat | wiederholen *v* | semifabricated | halbfertig *adj* |
| repeatability | Wiederhol- | sequence | Reihenfolge *f* |
| | genauigkeit *f* | service | Dienstleistung *f* |

| | | | |
|---|---|---|---|
| severe | streng *adj* | spline shaft | Keilwelle *f* |
| shaft | Welle *f* | spring | Feder *f* |
| shape | Form *f* | sprocket | Kettenrad *n* |
| shear | scheren *v* | spur gear | Stirnrad *n* |
| shear stress | Scher- | squeeze | auspressen *v* |
| | beanspruchung *f* | stability | Stabilität *f* |
| shears | Scherer *m* | stage | Phase *f* |
| shortcoming | Unzulänglichkeit *f* | stiffness | Steifheit *f* |
| shot or grit | körnchenblasen *v* | store | lagern *v* |
| blast | | strain | Verformung *f* |
| shot peen | kugelstrahlen,ver- | strain harden | kalthärten *v* |
| | festigungsstrahlen *v* | stream of air | Luftstrom *m* |
| sideways | seitwärts *adv* | strength | Festigkeit *f*, Stärke *f* |
| similar | ähnlich *adj* | stress | Spannung *f* |
| single point tool | Schneidmeißel *m* | stress relieve | spannungsarm |
| size | Größe *f* | | glühen *v* |
| skill | Geschicklichkeit *f* | stretch form | streckziehen *v* |
| slab | Platte *f* | strict | streng *adj* |
| sleeve | Muffe *f* | stringent | streng *adj* |
| sliding | gleitend *adj* | strip | Streifen *m* |
| slit | schlitzen *v* | stripper plate | Abstreifer *m* |
| slope | Steigung *f* | stroke | Hub *m* |
| slot | Schlitz *m* | stud | Stift *m* |
| smooth | zügig *adj* | sturdy | stark *adj* |
| socket spanner | Schrauben- | subject to | belasten mit *v* |
| | schlüssel *m* | successively | hintereinander *adv* |
| soften | weich machen *v* | suck | saugen *v* |
| solve | lösen *v* | sufficient | ausreichend *adj* |
| spark plug | Zündkerze *f* | suitable | geeignet *adj* |
| specification | Spezifizierung *f* | superfinishing | kurzhubhonen *v* |
| specimen | Versuchsgegen- | support | Träger *m* |
| | stand *m*, Exemplar *n* | surface finish | Oberflächenqualität *f* |
| spelt | Messinghartlot *n* | swivel | schwenken, drehen *v* |
| spin | drehen *v* | tailstock | Reitstock *m* |
| spindle | Spindel *f* | tang | Kegellappen *m* |

| | | | |
|---|---|---|---|
| **tap** | Gewindebohrer *m* | **valve** | Ventil *n* |
| **taper** | Verjüngung *f* | **vapour** | Dampf *m* |
| **taper shank** | Kegelschaft *m* | **variation** | Veränderung *f* |
| **tax** | Steuer *f* | **vernier caliper** | Messchieber *m* |
| **temper** | anlassen *v* | **versatile** | vielseitig *adj* |
| **tensional load** | zügige Belastung *f* | **vice** | Schraubstock *m* |
| **thermoform** | warmumformen *v* | **view** | Aussicht *f*, Blick *m* |
| **tight** | dicht, eng *adj* | **washer** | Unterlegscheibe *f* |
| **tip** | Spitze *f* | **wear** | Verschleiß *m* |
| **tolerance** | Toleranz *f* | **wedge** | Keil *m* |
| **tool post** | Meißelhalter *m* | **weld** | schweißen *v* |
| **torque** | Drehmoment *n* | **wide range** | große Auswahl *f* |
| **toughness** | Zähigkeit *f* | **work harden** | kalthärten *v* |
| **track** | verfolgen *v* | **work envelope** | Arbeitsraum *m* |
| **transform** | umwandeln *v* | **or work zone** | |
| **transition fit** | Übergangspassung *f* | **worm gear** | Schneckengetriebe *n* |
| **transverse** | quer *adj* | **wrinkle** | Falte *f* |
| **trim** | trimmen *v* | **wrist** | Handgelenk *n* |
| **tube** | Rohr *n* | **wrought iron** | Schmiedeeisen *n* |
| **turn** | drehen *v* | | |
| **turning centre** | CNC | | |
| | Drehmaschine *f* | | |
| **turret** | Revolverkopf *m* | | |
| **ultimate** | Zugfestigkeit *f* | | |
| **tensile strength** | | | |
| **ultimate** | endgültig *adj* | | |
| **undergo** | erfahren *v* | | |
| **unilateral** | einseitig *adj* | | |
| **unit** | Einheit *f* | | |
| **unite** | vereinigen *v* | | |
| **universal joint** | Gelenkkupplung *f* | | |
| **utilization** | Nutzung *f* | | |
| **utilize** | verwenden *v* | | |
| **vacuum** | Vakuum *n* | | |
| **value** | Wert *m* | | |

# Vocabulary 2 Deutsch / Englisch

| | |
|---|---|
| **abhängen** *v* | depend |
| **Abrieb** *m* | abrasion |
| **abschrecken** *v* | quench |
| **Abstreifer** *m* | stripper plate |
| **Abteilung** *f* | department |
| **abwärts** *adv* | downwards |
| **Abweichung** *f* | deviation |
| **Achse** *f* | axis |
| **ähnlich** *adj* | similar |
| **allmählich** *adv* | gradually |
| **Änderung** *f* | change |
| **Andeutung** *f* | indication |
| **angelernt** *adj* | semi-skilled |
| **Anlage** *f* | construction |
| **Anlagekosten** *pl* | fixed costs |
| **anlassen** *v* | temper |
| **annullieren** *v* | nullify |
| **anordnen** *v* | coordinate |
| **Anordnung** *f* | layout |
| **ansteigen** *v* | ascend |
| **Antrieb** *m* | actuator |
| **Anwendung** *f* | application |
| **Anzeige** *f* | indication |
| **anziehend** *adj* | attractive |
| **appellieren** *v* | appeal |
| **Arbeitskosten** *pl* | labour costs |
| **Arbeitsorgan** *n* | end effector |
| **Arbeitsraum** *m* | work zone or work envelope |
| **Arbeitsterminplan** *m* | schedule |
| **aufdrängen** *v* | impose |
| **aufeinanderwirken** *v* | interact |
| **Aufgabe** *f* | assignment |
| **aufkohlen** *v* | carburize |
| **Auflösung** *f* | resolution |
| **Aufnehmer** *m* | adaptor |
| **Auftrag erteilen** *v* | release |
| **ausklinken** *v* | notch |
| **auspressen** *v* | squeeze |
| **Auspuff** *m* | exhaust |
| **ausreichend** *adj* | sufficient |
| **Ausrichtung** *f* | alignment |
| **ausschalten** *v* | disengage |
| **ausschneiden** *v* | extract |
| **äußerlich** *adj* | external |
| **Aussicht** *f*, **Blick** *m* | view |
| **Ausstattung** *f* | equipment |
| **Austreiber** *m* | drift |
| **Auswahl** *f* | range |
| **auswählen** *v* | select |
| **auswechselbar** *adj* | interchangeable |
| **auswerfen** *v* | eject |
| **Automatenstahl** *m* | free-cutting steel |
| **automatisch** *adj* | automatic |
| **Bahnsteuerung** *f* | continuous path control |
| **Baukasten** *m* | module |
| **Baukasten-system** *n* | modular system |
| **Baustahl** *m* | mild steel |
| **Bearbeitungszugabe** *f* | machining allowance |
| **Befehl** *m* | instruction, order |
| **begründet auf** *v* | based on |
| **beherrschbar** *adj* | manageable |
| **Beispiel** *n* | example |

| Deutsch | Englisch |
|---|---|
| belasten mit *v* | subject to |
| beliebt *adj* | popular |
| Benzinpumpe *f* | fuel pump |
| berechnen *v* | evaluate |
| Bereich *m* | extent |
| Bericht erstatten *v* | report |
| beschäftigen *v* | employ |
| Beschreibung *f* | description |
| beseitigen *v* | eliminate |
| besorgen *v* | provide |
| Bestandsverzeichnis *n* | inventory |
| bestehen aus *v* | consist of |
| bestimmen | determine |
| betroffen *v* | concerned |
| Bettführungen *pl* | guideways |
| Bevorzugung *f* | preference |
| beweglich *adj* | flexible |
| Bewegung *f* | motion |
| bewerten *v* | assess, evaluate |
| biegen *v* | bend |
| billig *adj* | cheap |
| Bindemittel *n* | bonding material |
| Blei *n* | lead |
| Bohrer *m* | drill |
| Bohrfutter *n* | chuck |
| bördeln, falzen *v* | bead |
| Brennstoff *m* | fuel |
| Bruchfestigkeit *f* | breaking strength |
| Buchführung *f* | accounting |
| CNC Drehmaschine *f* | turning centre |
| Dampf *m* | vapour |
| Dauerfestigkeit *f* | fatigue strength |
| Defekt *m*, Fehler *m* | flaw |
| dehnbar, biegbar *adj* | ductile |
| Dehnbarkeit *f* | ductility |
| Dehnung *f* | strain |
| Deterministische Bedarfsermittlung *f* | material requirements planning (MRP) |
| dicht, eng *adj* | tight |
| Dienstleistung *f* | service |
| Dorn *m* | mandrel |
| drehbeweglich *adj* | rotary, rotatory |
| Drehkopf *m* | turret |
| Drehmaschine *f* | lathe |
| Drehmoment *n* | torque, moment |
| Dreibacken-bohrfutter *n* | self-centering chuck |
| Druck *m* | pressure |
| drücken, pressen *v* | press |
| Dübel *m* | dowel pin |
| durchführen *v* | execute |
| Eigenart *f* | character |
| Eigenschaft *f* | property, feature |
| Eindruck *m* | indentation |
| einführen *v* | introduce |
| Einführung *f* | introduction |
| Einheit *f* | unit |
| Einlaßventil *n* | inlet valve |
| einpressen *v* | force into |
| Einrichtung *f* | equipment |
| einsatzhärten *v* | case harden |
| einschließen *v* | include |
| einschneiden *v* | lance |
| Einschnitt *m* | recess |
| einseitig *adj* | unilateral |
| Einspritzdüse *f* | injection nozzle |
| Einzelheiten *pl* | details |
| eisenhaltig *adj* | ferrous |

| | | | |
|---|---|---|---|
| Eisenbeton *m* | reinforced concrete | flüssiges Metall *n* | molten metal |
| elastische Grenze *f* | elastic limit | Flüssigkeit *f* | fluid |
| endgültig *adj* | ultimate | Flußmittel *n* | flux |
| Endverfahren *n* | finishing process | Folgeschneid-werkzeug *n* | follow-on die set |
| Energie *f* | energy | Förderanlage *f* | transfer line |
| entfernen *v* | remove | Fördertechnik *f* | material handling |
| entgegen dem Uhrzeigersinn | anticlockwise | Form *f* | shape, form |
| enthalten *v* | contain | formen, bilden *v* | form |
| entscheiden *v* | decide | fortgeschritten *adj* | advanced |
| entsprechend *adj* | corresponding | Fortschritt *m* | progress |
| entwerfen *v* | design | Fräserdorn *m* | arbor |
| Entwicklung *f* | development | Freiheitsgrad *m* | degrees of freedom |
| entzünden *v* | ignite | | |
| erfahren *v* | undergo | Freiwinkel *m* | clearance angle |
| Erfahrung *f* | experience | Füller *m* | filler |
| erfordern, brauchen *v* | require | Futter *n* | chuck |
| erleichtern *v* | facilitate | Gefüge *n* | grain structure |
| ermitteln *v* | investigate | geeignet *adj* | suitable |
| erwägen, überlegen *v* | consider | Geldmittel *pl* | financial resources |
| erwähnen *v* | mention | | |
| erwerben *v* | acquire | Gelenk *n* | joint |
| erzeugen *v* | generate, create | Gelenkkupplung *f* | universal joint |
| existieren *v* | exist | genau *adj* | accurate |
| Fähigkeit *f* | ability | Genauigkeit *f* | accuracy |
| Falte *f* | wrinkle | gepreßtes Metallstück | metal pressing |
| Feder *f* | spring | | |
| feinkörnig *adj* | fine grain | Gerät *n* | device |
| Festigkeit *f*, Stärke *f* | strength | Geräusch *n*, Lärm *m* | noise |
| verfestigungs-strahlen *v* | shot peen | Geschicklichkeit *f* | skill |
| | | geschlossener Kreis *m* | closed loop |
| Fett *n* | grease | | |
| Flankenspiel *n* | backlash | Gesenkschmieden *n* | drop forging |
| fließen *v* | flow | Gestell *n* | frame |

| | | | |
|---|---|---|---|
| Gewinde *n* | screw thread | identisch *adj* | identical |
| Gewindebohrer *m* | tap | im Uhrzeigersinn | clockwise |
| Gewohnheit *f* | practice | ineinandergreifen *v* | interlock |
| gießen *v* | cast | Inhalt *m* | content |
| Gießform *f*, Form *f* | mould | innerlich *adj* | internal |
| glanzlos *adj* | matt | instand halten *v* | maintain |
| Gleichgewicht *n* | equilibrium | Instandhaltung *f* | maintenance |
| gleitend *adj* | sliding | kalthärten *v* | strain harden, |
| Glied *n* | link | | work harden |
| gravieren *v* | engrave | Kamm *m* | ridge |
| greifen *v* | grip | Kammer *f* | chamber |
| Greifer *m* | gripper | kaufen *v* | purchase |
| Grenze *f* | limit | Kegel *m* | cone |
| grobkörnig *adj* | coarse grain | kegelförmig *adj* | conical |
| Größe *f* | quantity, size | Kegellappen *m* | tang |
| Größenordnung *f* | magnitude | Kegelrad *n* | bevel gear |
| Grundbestandteil *m* | element | Kegelschaft *m* | taper shank |
| Guss *m* | casting | Kegelsenkung *f* | countersink |
| Gussblock *m* | ingot | Keil *m*, Paßfeder *f* | key |
| Gusseisen *n* | cast iron | Keilwelle *f* | spline shaft |
| halbfertig *adj* | semifabricated | Kenntnis *f*, Wissen *n* | knowledge |
| Hälfte *f* | Half (pl.halves) | Kerbe *f* | notch |
| Handgelenk *n* | wrist | Kette *f* | chain |
| Härte *f* | hardness | Kettenrad *n* | sprocket or |
| hartlöten *v* | braze | | sprocket wheel |
| Harz *n* | resin | Klemme *f* | clamp |
| herkommlich *adj* | conventional | Kohlenstoffstahl *m* | carbon steel |
| herstellen *v* | manufacture | kompensieren *v* | compensate |
| hintereinander *adv* | successively | komplex *adj* | complex |
| hinweisen *v* | refer | Konkurrenz *f* | competition |
| Hitze *f*, Wärme *f* | heat | Konsole *f*, Knie *n* | knee |
| hochglanzpolieren *v* | buff | Körnchen *n* | grains, granules |
| Hohlraum *m* | cavity | körnchenblasen *v* | shot or grit blast |
| honen *v* | hone | korrosiv, ätzbar *adj* | corrosive |
| Hub *m* | stroke | Kosten *pl* | costs |

| | | | |
|---|---|---|---|
| Kostengrenze *f* | financial limits | mangeln, fehlen *v* | lack |
| Kraft *f* | force | Maß *n* | dimension |
| kriechen *v* | creep | massivumformen *v* | bulk deform |
| Kriterium *n*, | criterion | Maßstab *m* | criterion |
| Maßstab *m* | | Mechanismus *m* | mechanism |
| kugelstrahlen, ver- | shot peen | Mehrschneidiges- | multipoint |
| festigungsstrahlen *v* | | werkzeug *n* | cutting tool |
| Kühlschmiermittel *n* | coolant | mehrzweck- *adj* | multipurpose |
| Kunde *m* | customer | Meisel *m* | chisel |
| Kupfer *n* | copper | Meißelhalter *m* | tool post |
| Kupplung *f* | coupling | Menge *f* | amount |
| Kurbel *f* | crank | messen *v* | measure |
| Kurbelwelle *f* | crankshaft | Messchieber *m* | vernier caliper |
| kurzhubhonen *v* | superfinishing | Messchraube *f* | micrometer |
| Lager *n* | bearing | Messuhr *f* | dial gauge |
| Lagerbuchse *f* | bush | Messung *f* | measurement |
| lagern *v* | store | Metallspäne *pl* | metal chips |
| längs *adj* | longitudinal | Methode *f*, Zugang *m* | approach, |
| läppen *v* | lap | | method |
| Last *f* | load | Mikrokristall *n* | crystal grain |
| Legierung *f* | alloy | Minimum *n* | minimum |
| leisten *v* | achieve, perform | Mischung *f* | mixture |
| Leistung *f* | power | montieren *v* | assemble |
| leistungsfähig *adj* | efficient | Motor *m* | engine, motor |
| Leistungsfähigkeit *f* | efficiency | Muffe *f* | sleeve |
| Leitspindel *f* | leadscrew | Muster *n* | prototype |
| Lichtbogen *m* | arc | Nabe *f* | hub |
| liefern *v* | deliver | Nachteil *m* | disadvantage |
| Lieferung *f* | delivery | nahtlos *adj* | seamless |
| linear *adj* | linear | Nennmass *n* | basic size |
| lockern *v* | loosen | netto *adj* | nett |
| Losdrehsicherung *f* | locking device | knabberschneiden *v* | nibble |
| Luftström *m* | stream of air | Niederhalter *m* | pressure pad |
| Machinestundensatz- | machine-hour | Niete *f* | rivet |
| rechnung *f* | rate | Nockenwelle *f* | camshaft |

| | | | |
|---|---|---|---|
| normalglühen *v* | normalize | Reibahle *f* | reamer |
| notwendig *adj* | necessary | Reibung *f* | friction |
| numerische | numerical | Reihe *f*, Bereich *m* | range |
| Steuerung *f* | control | Reihenfolge *f* | sequence |
| Nut *f* | groove, keyway | reinigen *v* | clean |
| Nutzung *f* | utilization | Rentabilitätsgrenze *f* | break-even |
| Oberflächenqualität *f* | surface finish | | number |
| obwohl *cj* | although | Reitstock *m* | taitstock |
| Ökologie *f* | ecology | Revolverdrehbank *f* | capstan lathe |
| passend, geeignet *adj* | suitable | Revolverkopf *m* | turret |
| pendelnde | reciprocating | Richtung *f* | direction |
| Bewegung *f* | motion | Riemen *m* | belt |
| Passung *f* | fit | Rißß *m*, Spalt *m* | crack |
| Phase *f* | stage | Roboterarm *m* | arm assembly |
| Platte *f* | slab | | or manipulator |
| Pleuelstange *f* | connecting rod | Rohr *n* | tube |
| prägen *v* | emboss | rückfeinen, | refine |
| Prägung *f* | impression | vergüten *v* | |
| prüfen *v* | inspect, check | Rückkopplung *f* | feedback |
| Punktsteuerung *f* | point to point | ruhig *adj* | quiet |
| | control | Rundstab *m* | rod |
| quer *adj* | transverse | Sach-, Fachkenntnis *f* | expertise |
| Querschnitt *m* | cross-section | saugen *v* | suck |
| Querschnitts- | necking | Säule *f* | column |
| verminderung *f* | | Säure *f* | acid |
| rau *adj* | rough | schaltbare | clutch |
| räumen *v* | broach | Kupplung *f* | |
| Rechnung stellen *v* | invoice | schätzen *v* | appraise |
| reduzieren , | reduce | Schätzung *f* | estimation |
| vermindern *v* | | Scherbeanspruch- | shear stress |
| Rechtecktisch *m* | compound table | ung *f* | |
| Regelkreis *m* | closed loop | Scherer *m* | shears |
| | control system | schieben *v* | push |
| regeln *v* | regulate | schleifen *v* | grind |

| | | | |
|---|---|---|---|
| Schleifmittel *n* | abrasive | sicherstellen *v* | ensure |
| Schleifscheibe *f* | abrasive wheel | Sollposition *f* | end point |
| Schlitz *m* | slot, slit | Spannung *f* | stress |
| Schrauben-schlüssel *m* | socket spanner | spannungsarm glühen *v* | stress relieve |
| Schmied *m* | blacksmith | Spannut *f* | flute |
| Schmiedeisen *n* | wrought iron | Spanwinkel *m* | rake angle |
| Schmierung *f* | lubrication | sparsam *adj* | economical |
| Schneckengetriebe *n* | worm gear | Spezifizierung *f* | specification |
| Schnecken-strangpresse *f* | screw feed mechanism | Spiel *n* | allowance |
| | | Spielraum *m* | clearance |
| Schneideisen *n* | die | Spindel *f* | spindle |
| Schneidkante *f* | cutting edge | Spindelstock *m* | headstock |
| Schneidmeißel *m* | single point tool | spinnen *v* | spin |
| Schneidplatte *f* | die block | Spitze *f* | tip |
| Schneidwerkzeug *n* | die set | spitzenlos schleifen *v* | centreless grind |
| schnell *adj* | rapid | Splint *m* | cotter pin |
| Schraube *f* | screw | Spritzguß *m* | die casting |
| Schraube *f* und Mutter *f* | bolt and nut | spröde *adj* | brittle |
| | | Sprödigkeit *f* | brittleness |
| Schraubenschlüssel *m* | socket spanner | stabil *adj* | rigid |
| Schraubstock *m* | vice | Stabilität *f* | stability |
| schrittweise *adj* | progressive | ständig *adj* | continuous |
| schwanken *v* | fluctuate | Stange *f*, Stab *m* | bar |
| schweißen *v* | weld | stanzen *v* | blank |
| schwenken, drehen *v* | swivel | stanzen, lochen *v* | punch |
| schwer *adj* | difficult | Stanzwerkzeug *n* | punch |
| schwer zugänglich *adj* | awkward | stark *adj* | sturdy |
| Schwungrad *n* | fly wheel | starre Kupplung *f* | rigid coupling |
| sehr *adv* | extremely | Steifheit *f* | stiffness |
| serienmäßig herstellen *v* | mass produce | Steigung *f* | lead |
| | | Steuer *f* | tax |
| streng, sorgfältig *adj* | rigorous | steuern *v* | open loop control |
| sich drehen *v* | revolve | | |
| sich vermindern *v* | decrease | Stift *m* | stud |

| | | | |
|---|---|---|---|
| **Stirnrad** *n* | spur gear | **undeutlich** *adj* | fuzzy |
| **Stoß** *m*, **Schlag** *m* | impact | **Universalteilkopf** *m* | dividing head |
| **Stößel** *m* | ram | **unmöglich** *adj* | impossible |
| **streckziehen** *v* | stretch form | **unter** *pr* | amongst |
| **Streifen** *m* | strip | **Unterlegscheibe** *f* | washer |
| **streng** *adj* | severe | **Unternehmen** *n* | enterprise |
| **streng, sorgfältig** *adj* | rigorous | **Unzulänglichkeit** *f* | shortcoming |
| **Teil** *m* | part | **Vakuum** *n* | vacuum |
| **Tiefe** *f* | depth | **Ventil** *n* | valve |
| **tiefziehen** *v* | deep draw | **Verabredung** *f* | appontment |
| **Toleranz** *f* | tolerance | **Veränderung** *f* | variation |
| **Träger** *m* | support | **verbessern** *v* | improve |
| **Tragfähigkeit** *f* | pay load | **verbinden** *v* | join |
| | capacity | **Verbindung** *f* | compound |
| **treiben** *v* | drive | **verbrennen** *v* | burn |
| **trimmen** *v* | trim | **Verbrennung** *f* | combustion |
| **Trommel** *f* | drum | **verbunden** *v* | associated |
| **trommelpolieren** *v* | barrel finishing | **Verfahren** *n* | process, |
| **Übereinstimmung** *f* | accordance | | procedure |
| **Übergangspassung** *f* | transition fit | **vereinigen** *v* | unite |
| **überlagern** *v* | interfere | **verfolgen** *v* | track |
| **überwachen** *v* | monitor | **Verformung** *f* | deformation |
| **übrigbleiben** *v* | remain | **Vergaser** *m* | carburettor |
| **Uhrzeigersinn** | clockwise | **vergüten** *v* | rcfine |
| **Umdrehung** *f* | revolution | **Verhalten** *n* | behaviour |
| **umfassend** *adj* | comprehensive | **Verhältnis** *n* | ratio |
| **Umgebung** *f* | environment | **verhindern** *v* | prevent |
| **Umkreis** *m* | circumference | **Verjüngung** *f* | taper |
| **umlaufend** *adj* | recirculation | **Verkauf** *m* | sale |
| **Umsetzung** *f* | conversion | **verlängern** *v* | elongate |
| **umwandeln** *v* | convert | **Verlängerung** *f* | elongation |
| **Umwelt-** | pollution | **vermindern** *v* | reduce |
| **verschmutzung** *f* | | **verringern** *v* | diminish, reduce |
| **Unterteil** *m* | base | **Verschleiß** *m* | wear |

| | | | |
|---|---|---|---|
| verstärken v | reinforce | Werkzeugmaschine f | machine tool |
| Versuchsgegen-stand m | specimen | Werkzeugschlitten m | carriage |
| | | Wert m | value |
| Verwaltung f | administration | wesentlich adj | essential |
| verweisen v | refer | wichtig adj | important |
| verwenden v | utilize | Widerstand m | resistance |
| verwickeln v | involve | widerstandsfähig adj | hardwearing |
| verwirklichen v | realize | wiederbewerten v | recycle |
| vielkristallin adj | polycrystalline | wiederholen v | repeat |
| vielseitig adj | versatile | Wiederhol-genauigkeit f | repeatability |
| vollkommen adj | absolutely | | |
| voraussagbar adj | predictable | wiederverwenden v | reuse |
| voraussagen v | forecast, predict | Wirkungsgrad m | efficiency |
| vorbereiten v | prepare | wünchenswert adj | desirable |
| vorbestimmen v | predetermine | Zähigkeit f | toughness |
| Vorgang m | process | Zahnrad n | gear (or gear wheel) |
| vorhanden adj | available | | |
| vorher adj | previous | Zahnräder mit | helical gears |
| Vorrichtung f | jig, fixture | Schrägverzahnung pl | |
| vorspringen v | project out | Zahnstange f | rack |
| vorspringender Teil | projection | Zahnstangen-getriebe n | rack and pinion |
| vorstellen v | conceive | Zugfestigkeit f | ultimate tensile strength |
| Vorteil m | advantage | | |
| vorwärtsbewegen v | advance | Zubehör n | accessory |
| wählen, aufsuchen v | choose | zügige Belastung f | tensional load |
| wahrscheinlich adv | probably | Zündkerze f | spark plug |
| Wärmebehandlung f | heat treatment | zunehmen v | increase |
| warmumformen v | thermoform | zurückkehren v | revert |
| wechseln v | change | zusammendrücken v | compress |
| weich machen v | soften | zusammengesetzt adj | composite |
| weichglühen v | anneal | zusammengesetzt auf | composed of |
| weiterleiten v | route | Zusatzgerät n | attachment |
| Welle f | shaft | Zuverlässigkeit f | reliability |
| Werkzeug n | tool | Zwang m, Nötigung f | constraint |

| | | | |
|---|---|---|---|
| **zwanglos** *adv* | freely | **Zylinderkopf** *m* | engine head |
| **Zweck** *m* | purpose | **zylindrische** | counterbore |
| **zwingen, nötigen** *v* | compel | **Senkung** *f* | |
| **Wiederhol-** | repeatability | | |
| **genauigkeit** *f* | | | |

# Index

# Appendices

## Appendix I. List of materials – Werkstoffe (see also Chapters 1 & 2 )

### (i) Iron, Iron ore, Cast iron  –  Eisen, Eisenerz, Gußeisen

| | | | |
|---|---|---|---|
| alloy cast irons | legierte Eisen-Gußwerkstoffe | spheroidal cast iron | Gußeisen mit Kugelgraphit |
| blackheart malleable cast iron | Scwarzer Temperguß | pig iron | Roheisen |
| grey cast iron | Grauguß | slag | Schlacke |
| iron ore | Eisenerz | vermicular | Gußeisen mit |
| malleable cast iron | Temperguß | graphite cast iron | Vermiculargrapit |
| meehanite | Meehanite | whiteheart malleable cast iron | weißer Temperguß |

### (ii) Steels - Stähle

| | | | |
|---|---|---|---|
| alloy steels | legierte Stähle | high speed steels | Schnellarbeitsstähle |
| carbon steel | Kohlenstoffstahl | mild steel | Baustahl |
| case hardening steels | Einsatzstähle | nitriding steels | Nitrierstähle |
| | | spring steels | Federstähle |
| cast steel | Stahlguß | stainless steels | Nichtrostende Stähle |
| free-cutting steel | Automatenstahl | tool steels | Werkzeugstähle |
| hardenable steels | Vergütungsstähle | | |

### (iii) Nonferrous metals - Nichteisenmetalle

| | | | |
|---|---|---|---|
| aluminium | Alunimium | lead | Blei |
| bismuth | Wismut | tantalum | Tantal |
| cadmium | Kadmium | tin | Zinn |
| copper | Kupfer | titanium | Titan |
| magnesium | Magnesium | zinc | Zink |
| nickel | Nickel | | |

### (iv) Nonferrous metal alloys - NE Metall Legierungen

| | | | |
|---|---|---|---|
| aluminium alloys | Aluminium Legierungen | magnesium alloys | Magnesium Legierungen |
| copper alloys | Kupfer Legierungen | nickel alloys | Nickel Legierungen |
| brass (Cu -Zn) | Messing | tin alloys | Zinn Legierungen |
| bronze (Cu – Zn) | Zinnbronze | titanium alloys | Titan Legierungen |
| lead alloys | Blei Legierungen | zinc alloys | Zink Legierungen |

## Appendix I. List of materials (continued) – Werkstoffe

### (v) Plastics - Kunstoffe

#### (a) Thermoplastics - Thermoplaste

| | | | |
|---|---|---|---|
| **polyamide (PA), nylon** | Polyamide | **Polypropylene (PP)** | Polypropylen |
| **polycarbonate (PC)** | Polycarbonat | **Polystyrene (PS)** | Polystyrol |
| **polyethylene (PE)** | Polyethylen | **Polytetrafluoro-** | Polytetrafluor- |
| **polymethylmethacrylate** | Polymethylmethacrylat, | **ethylene (PTFE)** | ethylen |
| **(PMMA), Plexiglas** | ( PMMA) Acrylglas | | |

#### (b) Thermosetting plastics – Duroplaste

| | | | |
|---|---|---|---|
| **epoxy resins (EP)** | Epoxidharze | **silicone resins (SI)** | Silikonharze |
| **melamine resins (MF)** | Melaminharze | **unsaturated** | ungesättigte |
| **phenolic resins (PF)** | Phenolharze | **polyester resins** | Polyesterharze |
| **polyurethane resins (PUR)** | Polyurethanharze | | |

#### (c) Elastomers – Elastomere

| | |
|---|---|
| **synthetic rubber** | Synthesekautschuk |

### (vi) Additives - Zusatzstoffe

| | | | |
|---|---|---|---|
| **accelerators** | Beschleuniger | **lubricants and** | Schmiermittel und |
| **antioxidants** | Oxydierungsschutz-mittel | **flow promoters** | Strömförderungsmittel |
| | | **plasticizers** | Plastifizierungs-mittel |
| **dies and pigments** | Farbstoffe und Pigmente | **solvents** | Lösungsmittel |

### (vii) Fillers - Füllstoffe

| | | | |
|---|---|---|---|
| **alumina** | Tonerde | **talc** | Talk |
| **clay** | Lehm | **wood flour** | Holzmehl |
| **minerals** | Mineralien | **synthetic fibers** | synthetische Fasern |
| **quartz** | Quartz | | |

### (viii) Composite materials - Verbundwerkstoffe

| | | | |
|---|---|---|---|
| **coated materials** | beschichtete Materialen | **particle reinforced materials** | teilchenverstärkte Verbundwerkstoffe |
| **carbon, glass or other fiber reinforced materials** | andere faserverstärkte Materialen | **plywood** | Sperrholz |
| | | **sintered materials** | Sinterwerkstoffe |
| **laminated materials** | Schichtverbund-werkstoffe | | |

## Appendix II. List of tools – Werkzeuge (see also chapters 3, 4, 16 & 17)

| | | | |
|---|---|---|---|
| bandsaw | Bandsäge | grinding wheel | Schleifscheibe |
| bench shears, | Tafelschere. | hacksaw | Bügelsäge |
| guillotine | Guillotine | hammer | Hammer |
| boring bar with | Bohrstange mit | hone | Honstein |
| inserted tip | Schneideinsatz | honing stone | Honahle |
| broach | Räumahle | lap | Läppwerkzeug |
| centre punch | Körner | lapping paste | Läppegemisch |
| chisel | Meißel | lathe turning tool | Drehmeißel |
| clamp | Klammer | mallett | Holzhammer |
| clamping device | Spannelement | milling cutter | Fräswerkzeug |
| compass | Spitzzirkel | pliers | Zange |
| counterboring tool | Zylindersinker | reamer | Reibahle |
| countersinking tool | Kegelsinker | ring spanner | Ringschlüssel |
| die (thread cutting) | Schneideisen | saw | Säge |
| die (in a die set) | Schneidplatte | screwdriver | Schraubendreher |
| die set | Schneidwerkzeug | scriber | Reißnadel |
| divider | Stechzirkel | scribing block | Parallelanreißer |
| drill | Bohrer | spanner | Schraubenschlüssel |
| drill ( taper shank) | Bohrer  (Kegelschaft) | (american: wrench) | |
| follow-on die set | Folgeschneidwerkzeug | tap | Gewindebohrer |

## Appendix III. List of measuring devices and instruments
### Meßgeräte und Meßinstrumente (see also Chapter 3)

| | | | |
|---|---|---|---|
| coordinate meas- | Koordinaten- | plug gauges | Grenzlehrdorne |
| uring instruments | meßgeräte | pneumatic meas- | pneumatische |
| dial indicator, dial | Meßuhr | uring instruments | Meßgeräte |
| gauge (Amer:  gage) | | protractor | Winkelmesser |
| electronic meas- | elektronische | ring gauges | Lehrringe |
| urung instruments | Meßgeräte | rule, ruler | Maßstab, Lineal |
| form  gauges, | Formlehren | sine bar | Sinuslineal |
| contour gages, | | slip gauges | Endmasse |
| templates | | (amer: gage blocks) | |
| gap gauges | Grenzlehren | spirit level | Richtwaage |
| inside calipers | Innentaster | surface plate | Meßplatte |
| inside micrometer | Innenmeßschraube | thread gauges | Gewindelehren |
| instruments for the | Geräte für die | vernier | Nonius |
| measurement of | Prüfung von | vernier caliper | Meßschieber |
| (a) roundness | (a) Rundheit | vernier depth gauge | Tiefenmeßschieber |
| (b) concentricity | (b) Konzentrität | vernier height | Höhenmeßschieber |
| (c) coaxiality | (c) Koaxialität | gauge | |
| limit gauges | Grenzlehren | vernier with a dial | Meßschieber mit |
| micrometer | Meßschraube | indicator | Rundscala |
| outside calipers | Außentaster | | |